Electrochemistry of Natural Compounds

Electrochemistry of Natural Compounds explores cutting-edge techniques in electrosynthesis, emphasizing green chemistry principles for the sustainable production of these valuable molecules. It delves into the application of natural compounds in nanoparticle biosynthesis and their subsequent electrochemical applications. A significant portion of the book focuses on the investigation of natural compounds as corrosion inhibitors, utilizing Tafel polarization and electrochemical impedance spectroscopy. Further, it provides a thorough review of voltammetric methods, including cyclic voltammetry, linear sweep voltammetry, differential pulse voltammetry, and more, highlighting electrode modification strategies and optimization techniques. Practical aspects, such as solution preparation, electrode fabrication, and the impact of pH, are addressed, making this a hands-on resource for researchers and students.

This book offers a unique combination of electrosynthesis, corrosion inhibition, and voltammetric measurement techniques specifically tailored for the analysis and application of natural compounds. This holistic approach fills a gap in current literature, providing researchers with a valuable resource for developing novel electrochemical methods for analyzing and utilizing natural products across diverse scientific disciplines.

Key Features:

- Introduces terms and concepts used to describe electrochemical reactions.
- Covers green chemistry, oxidation, and reduction early to provide a foundation for understanding later topics.
- Explores factors that affect voltammetric methods used to determine natural compounds.
- Addresses coverage of green nanoparticle synthesis using these methods.
- Studies the use of cyclic voltammetry (CV), linear sweep voltammetry (LSV), chronocoulometry, differential pulse voltammetry (DPV), and rotating disc electrode (RDE).

Electrochemistry of Natural Compounds

Maryam Khayatkashani, Seyed Mohammad Nabavi,
Seyed Fazel Nabavi, Ana Sanches Silva, and
Farheen Badrealam Khan

CRC Press
Taylor & Francis Group
Boca Raton London New York

CRC Press is an imprint of the
Taylor & Francis Group, an **informa** business

First edition published 2026
by CRC Press
2385 NW Executive Center Drive, Suite 320, Boca Raton FL 33431

and by CRC Press
4 Park Square, Milton Park, Abingdon, Oxon, OX14 4RN

CRC Press is an imprint of Taylor & Francis Group, LLC

ISBN: 978-1-032-13400-0 (hbk)
ISBN: 978-1-032-13406-2 (pbk)
ISBN: 978-1-003-22903-2 (ebk)

DOI: 10.1201/9781003229032

Typeset in Times LT Std
by Apex CoVantage, LLC

Dedicated to:

To my father, Abbas Khayatkashani, and my mother, Fatemeh Kasiri, for their endless love and support, and to my daughter, Asal, the brightest star in my sky.
Maryam Khayatkashani

With love and in memory of Seyed Ali Asghar Nabavi and Fatemeh Soghra Norouznejad, we dedicate this book to our mother, Maryam Nabavi, whose unwavering support has made this possible.
Seyed Fazel Nabavi and Seyed Mohammad Nabavi

For my parents, Maria Lucília and Jacinto, whose unwavering love and endless support are the foundation of everything.
Ana Sanches Silva

To our beloved mother, Khaleeda Parveen; sister, Aisha Abrar Shaikh; brother, Mohammed Atif Khan; husband, Mirza Shahrukh Baig; uncle and aunts; and our lovely nieces and nephews – whose unwavering support has made this possible.
Farheen Badrealam Khan

Contents

Preface

For centuries, humanity has turned to nature for sustenance, healing, and inspiration. Within the intricate structures of natural compounds lie solutions to some of our most pressing challenges, from the development of novel pharmaceuticals to the creation of sustainable materials. Increasingly, electrochemistry is emerging as a powerful and versatile tool for unlocking the potential of these bioactive compounds.

This book, *Electrochemistry of Natural Compounds*, aims to bridge the gap between the rich field of natural products chemistry and the precise control offered by electroanalytical techniques. It reflects the growing recognition that electrochemistry is not merely a method of analysis but also a means of synthesis, a way to understand fundamental mechanisms, and a driver of green technologies.

This book is provided a comprehensive exploration of the field. Initially, we review the electrosynthesis of natural compounds, emphasizing green chemistry principles through oxidation and reduction strategies. We then explore nanoparticle biosynthesis and its exciting applications within electrochemistry. Following this, we examine the protective capabilities of nature, focusing on natural compounds as corrosion inhibitors, using techniques like Tafel polarization and electrochemical impedance spectroscopy. A significant portion of the book is dedicated to the measurement of natural compounds using voltammetric methods, covering various techniques from cyclic voltammetry to rotating ring-disc electrodes, electrode modification with carbon paste and glassy carbon electrodes, the influence of reactive functional groups, solution preparation, electrode manufacture, and optimization strategies. Finally, we review practical aspects like measurements in real samples and simultaneous compounds determination.

We begin by exploring the electrosynthesis of natural compounds, embracing the principles of green chemistry to offer sustainable and efficient routes to complex molecules through oxidation and reduction reactions. We then analyze the fascinating realm of nanoparticle biosynthesis, demonstrating how nature can be harnessed to create materials with unique electrochemical properties and applications.

Beyond synthesis, we consider the protective power of nature. The book examines the role of natural products as corrosion inhibitors, detailing the use of Tafel polarization and electrochemical impedance spectroscopy to understand their effectiveness in safeguarding materials.

A significant portion of our book is dedicated to the measurement of natural compounds using voltammetric methods. We provide a comprehensive overview of these techniques, from the fundamentals of cyclic voltammetry to the intricacies of differential pulse voltammetry. We explore the recent advancements in electrode modification, showcasing how carefully selected modifiers can enhance sensitivity and selectivity. Practical guidance is offered on solution preparation, electrode fabrication, and the optimization of key experimental parameters. Finally, we review the challenges of analyzing real-world samples and the exciting possibilities of simultaneous compound determination.

This book is intended for a broad audience, including graduate students, researchers, and professionals working in chemistry, biology, materials science, and related fields. It is our hope that it will serve as a valuable resource, stimulating new ideas and fostering collaboration between those who seek to understand and harness the power of electrochemistry in the service of a more sustainable and healthier future. We are deeply grateful to the contributors who have shared their expertise and insights, and to CRC Press for their support.

Maryam Khayatkashani
Seyed Fazel Nabavi
Ana Sanches Silva
Farheen Badrealam Khan
Seyed Mohammad Nabavi

Editor Biographies

Maryam Khayatkashani is a senior scientist at Advanced Medical Pharma (AMP-Biotec), Biopharmaceutical Innovation Centre, Via Cortenocera, 82030, San Salvatore Telesino (BN), Italy. Her research focuses on bioelectrochemistry, biosynthesis of nanoparticles, computational chemistry, and natural products. She is the author/co-author of numerous publications in international journals and books and has presented at numerous national and international conferences.

Seyed Mohammad Nabavi is a senior scientist at Advanced Medical Pharma (AMP-Biotec), Biopharmaceutical Innovation Centre, Via Cortenocera, 82030, San Salvatore Telesino (BN), Italy. His research focuses on natural products. He is the author of over 12 books and more than 400 publications in ISI-indexed peer-reviewed journals. His work has garnered over 30,000 citations. Clarivate's Essential Science Indicators lists Seyed Mohammad among the top 1% of scientists worldwide in agricultural sciences and pharmacology and toxicology.

Seyed Fazel Nabavi is a senior scientist at Advanced Medical Pharma (AMP-Biotec), Biopharmaceutical Innovation Centre, Via Cortenocera, 82030, San Salvatore Telesino (BN), Italy. His research focuses on the health-promoting effects of natural products. He is the author/co-author of over 250 publications in international journals, more than 5 books, and 4 book chapters. He has presented at over 45 national and international conferences and serves as an editorial board member and reviewer for several international journals. Clarivate's Essential Science Indicators lists Seyed Fazel among the top 1% of scientists worldwide in agricultural sciences and pharmacology and toxicology.

Ana Sanches Silva is a professor at the Faculty of Pharmacy, University of Coimbra (Portugal). Ana holds a degree in Pharmaceutical Sciences from the University of Coimbra and earned her European Ph.D. in Pharmacy with honors from the University of Santiago de Compostela, Spain. Ana Sanches Silva has a remarkable track record with more than 300 scientific publications. Her research focuses on the development of innovative packaging technologies, such as active and edible packaging, and the study of packaging migration into food. She has a strong interest in analytical methodologies for detecting bioactive compounds, food contaminants (e.g. plant toxins, mycotoxins, pesticides residues), and packaging migrants. In recent years, she has also explored natural extracts from plants and food by-products. Throughout her distinguished career, Ana has received over 20 research awards, recognizing her outstanding contributions to science. Additionally, Ana has been consistently recognized among the World's Top 2% of most-cited scientists in the Stanford University ranking since 2021.

Farheen Badrealam Khan is a postdoctoral fellow in the Department of Biological Sciences, College of Medicine and Health Sciences, Khalifa University, Abu Dhabi, UAE. She holds a PhD in biotechnology and is a member of the Royal Society of Biology. Her research focuses on investigating the therapeutic potential of bioactive molecules, nanoparticles, and peptides against diverse disease pathologies. She has authored and co-authored numerous publications in international journals and books and has presented her work at various national and international conferences.

1 Electrosynthesis of Natural Compounds

Pardis Mohammadi Pour[1, 2], *Atif Ali Khan Khalil*[3], *and Mi-jeong Ahn*[4]

[1] Food and Drug Research Center, Food and Drug Administration, Ministry of Health and Medical Education, Tehran, Iran.

[2] Department of Pharmacognosy, Faculty of Pharmacy, Tehran University of Medical Sciences, Tehran, Iran.

[3] Department of Biotechnology, Yeungnam University, Gyeongsan, Republic of Korea.

[4] College of Pharmacy and Research Institute of Pharmaceutical Sciences, Gyeongsang National University, Jinhu, Republic of Korea.

1.1 INTRODUCTION

The synthesis of natural products as a part of synthetic organic chemistry has remained largely unchanged over time and requires further development of more efficient methods [1, 2]. In the past few years, radical substitution reactions have engrossed special attention in the guise of suggesting the replacement of retrosynthetic disconnections of target compounds via one-electron, open-shell synthons in comparison with classical intermediates with two-electron closed shell [1, 3]. Open-shell intermediates can be produced by redox-active reagents, either in their reduced or oxidized states, and as well as by applying electrochemical or photochemical methods [4, 5]. Further, with the application of more sustainable processes and approaches, and with an improved atom economy, the assembly of intricate scaffolds can be realized [1, 6].

Organic electrosynthesis is currently recognized by the chemical society as a powerful yet not adequately developed synthetic procedure [7]. In an ordinary electrosynthesis chamber, two electrodes, an anode and a cathode, are plunged into a liquid environment comprised of a substrate and the requisite electrolyte necessary to increase the electrical conductivity of the liquid environment (Figure 1.1). Using potential difference between two electrodes leads to both oxidation and reduction reactions, assisted by an external power source. Electrochemical procedures offer a substantial advantage in comparison to the use of stoichiometric chemical oxidizing/reducing compounds. The highly reactive and/or toxic agents in an electrosynthesis chamber can either be completely avoided (in a direct approach) or produced in situ

DOI: 10.1201/9781003229032-1

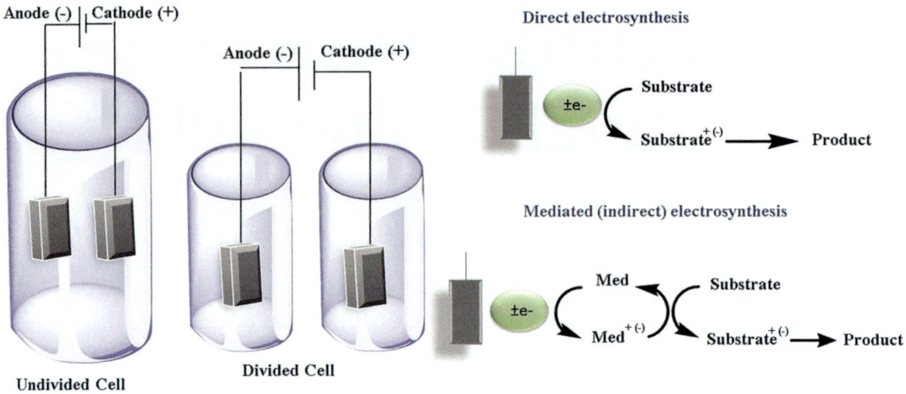

FIGURE 1.1 Illustration of a divided and undivided electrosynthetic cells (*left*). Typical schematic of direct and indirect electrosynthesis (*right*).

on the surface of the electrode in a controlled manner (in a mediated approach) [8]. Regulating the cell potential helps prevent undesirable side reactions, including overreduction and/or overoxidation, challenges that can be avoided when the stoichiometric content of redox materials is used. Oxidation (which occurs in the anodic part) or reduction (which occurs in the cathodic part) can be carried out in an undivided cell, and subsequently, more attention must be paid to possible interference in the auxiliary electrode, which may affect the desired transformation at the working electrode. If such interference inhibits the desired reaction, a divided cell can be used, as both the anodic and cathodic compartments are separated by an ion-exchange or porous membrane [1].

Electrosynthesis is usually conducted under either constant current or constant potential conditions. When constant current is used, the rate of the redox reaction is established and the cell potential adapts to a value necessary for driving the oxidation and reduction half-reactions at the stated rate. In this two-electrode setup, the potential at each electrode is determined by the thermodynamic redox potentials of electroactive species, kinetics of heterogeneous electron transfer (between electrode surface and solution species), and mass transport at the electrode surface. Therefore, the current should be chosen to be sufficiently large to surpass the heterogeneous charge-transfer resistance of the desired redox process but not exceed mass transport limit. If it does, too extreme values in cell potential will result, leading to destructive side reactions, like the oxidation/reduction of supporting electrolyte, solvent, or electrode material, thus reducing overall current efficiency.

During constant cell current electrolysis with a three-electrode setup, the cell potential gradually increases as the target electroactive species are consumed, and thus, the end point of the electrolysis should be carefully chosen to maximize substrate conversion without excessively increasing the cell potential. In contrast, constant potential electrolysis prevents side reactions at high substrate conversions but may limit the rate of the desired transformation. Additionally, in this case, one must consider and account for the potential drop between the working and reference electrodes due to solution resistance (iR-drop), which can be determined through analytical electrochemical measurements of the target electroactive species.

Until recently, the widespread use of organic electrosynthesis in synthetic organic chemistry has been restricted owing to the absence of standardized electrosynthetic cells and electrodes, resulting in poor reproducibility [9, 10]. In addition, most of the literature on electrochemically driven transformations of organic constituents concentrate on specific reactions of substrates which have a single electroactive functional system [11]. Despite the fact that a number of these reactions have attracted commercial applications [6], a more systematic procedure for showing the utility of organic electrosynthesis is required, including precise studies on substrate scope and functional group tolerance. In this chapter, selected examples demonstrate the successful application of organic electrosynthesis in creating complex molecules, with great attention to the developments, along with a couple of early illustrative examples [6]. Overall, with an increasing demand for sustainable and efficient synthetic methods, organic electrosynthesis is poised to become an essential tool in synthetic organic laboratories, offering a range of benefits in terms of safety, selectivity, and scalability.

1.2 GREEN CHEMISTRY

Redox reactions, which involve the transfer of electrons, are crucial in chemical science and are commonly studied and utilized in synthetic organic chemistry. Traditionally, to achieve oxidation or reduction reactions, large amounts of oxidizing or reducing agents are required [7, 12–16]. While these methods are effective in creating the desired compounds, the requirement of precise quantities of oxidizing or reducing agents leads to the production of dangerous waste, causing environmental issues in the implementation and expansion of these techniques. Organic electrosynthesis is a potent and efficient method of synthesis that can achieve redox changes by means of anodic oxidation and cathodic reduction, without requiring the presence of external oxidants or reducing agents [17]. When it comes to avoiding the utilization of precise amounts of oxidants or reducing agents, organic electrosynthesis is unequivocally a more environmentally friendly synthetic approach in contrast to conventional methods. An electrochemical C-H amination technique utilizing aromatic amides and cyclic secondary amines was revealed by Ackermann et al. in 2018 using a renewable solvent (tetrahydro-2H-pyran-2-one; Figure 1.1a) [18]. The Lei research team also published a divided cell co-catalyzed electrochemical C-H amination process between aromatic amides and secondary amines simultaneously (Figure 1.1b) [19]. The use of the cobalt catalyst and electrochemical methods facilitated the direct recycling of the catalyst via anodic oxidation, eliminating the need for 2.5 equivalent of $AgNO_3$ as an oxidant commonly used in traditional C-H amination methods (Figure 1.1c) [20]. As a result, this cleaner protocol offered an eco-friendlier approach to C-H amination while avoiding the production of waste products caused by oxidants (Figure 1.1c).

1.3 WASTE PREVENTION

With the fast growth of industrialization in the current economic model, the environmental implications of chemical production have become a greater concern. In addressing this concern, researchers are seeking ways to reduce, and ideally eliminate, waste generation. To address the issue of waste generation during chemical production, chemists are placing greater emphasis on developing reactions that occur under waste-free conditions. One promising and environmentally friendly approach for bond formation is the electrochemical oxidative cross-coupling of R_1-H with R_2-H [21]. This approach enables the production of the desired cross-coupling product through direct or indirect anodic oxidation under either metal or metal-free conditions. In addition, it produces valuable H_2 gas as a by-product through the cathodic proton reduction process (Figure 1.2). Electrochemical oxidative cross-coupling of R_1-H with R_2-H offers a more environmentally friendly alternative to traditional oxidative cross-coupling methods that typically result in the generation of significant waste. This is because the former approach leads to the same product without the production of waste, making it a far greener option (Figure 1.2) [21].

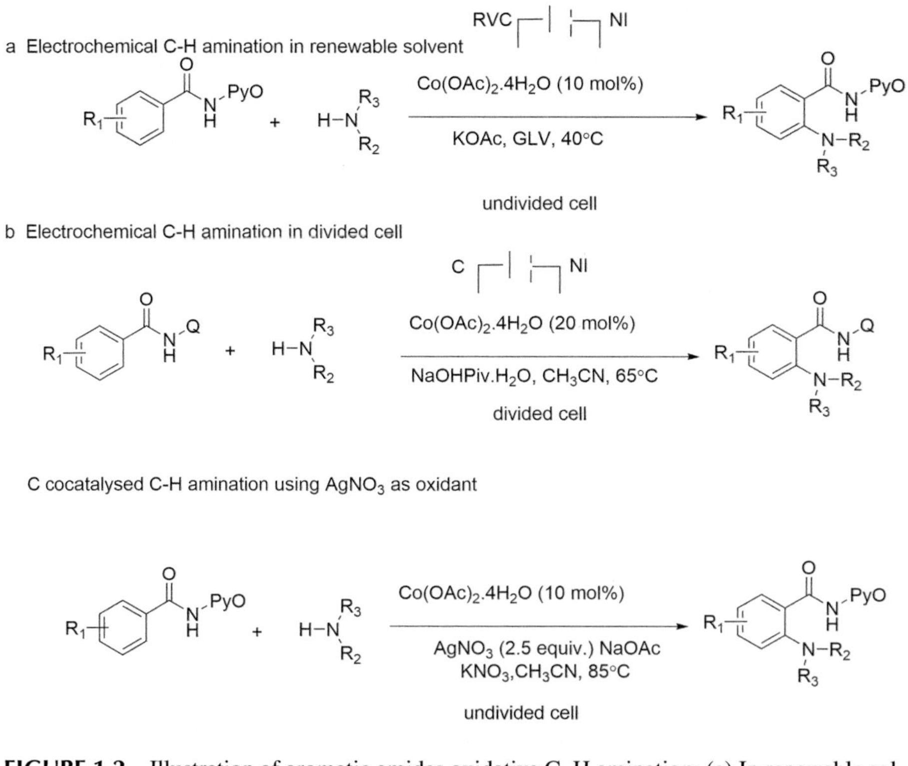

FIGURE 1.2 Illustration of aromatic amides oxidative C–H amination: (a) In renewable solvent. (b) In divided cell. (c) By applying $AgNO_3$ as oxidant.

1.4 MORE ADVANTAGES OF ELECTROSYNTHESIS COMPARED TO TRADITIONAL METHODS

The environmental advantages of electrosynthesis are merely one of the motivating factors that should prompt scientists to adopt this technique [22]. Electrochemical reactions are usually very adaptable to functional groups in the absence of external oxidants or reductants. These reactions can be performed at lower temperatures and pressures than conventional methods, making them more energy-efficient [22]. In addition, the time required to carry out these reactions is usually brief because of their efficient reaction rate, and this time can also be decreased with an increase in the operating current. Moreover, the electrochemical system's oxidation or reduction potential can be modified by adjusting the current or voltage, allowing for reactions that cannot perform by using chemical oxidants or reductants [22]. It should be noted that most electrochemical reactions can be halted quickly by just turning off the power switch, unlike conventional reactions that usually necessitate quenching. Lastly, electrosynthesis is highly promising for industrial uses because it can be easily expanded for larger-scale production. This method has already been successfully industrialized for producing 1,4-dicyanobutane and sebacic acid [22].

1.5 NOT ALL ELECTROCHEMICAL REACTIONS ARE GREEN

The term *green chemistry* refers to the creation of chemical products and procedures with the goal of minimizing or eliminating the application and production of hazardous substances [23]. The "12 Principles of Green Chemistry" [24] provides various measures to evaluate whether a reaction is environmentally friendly, or to compare the ecological effects of different reactions, by evaluating specific metrics. Just because stoichiometric quantities of oxidants or reductants are utilized and/or H_2 gas is produced during electrochemical reactions doesn't automatically guarantee that these reactions are environmentally friendly. Taking into consideration all principles of green chemistry enable a comprehensive evaluation of the environmental influence of a chemical transformation. In particular, hazardous electrolytes are commonly utilized in organic electrosynthesis as an additive for increased conductivity in solution; nevertheless, this increases the risk of explosion. Additionally, toxic, flammable, or corrosive solvents might be selected for particular electrochemical processes, which can have a negative effect on the safety aspects of the process. Environmentally unfriendly additives can also be implemented in electrosynthesis and cause a substantial amount of waste to be produced. In all such cases, electro-organic synthesis does not adhere to the fundamental principles of green chemistry [12].

In the last decade, organic electrosynthesis has been employed as an alternative to toxic and hazardous oxidizing or reducing reagents in order to reduce energy consumption and facilitate the in-situ production of unstable and hazardous reagents. This study provides an overview of electrosynthetic processes, including paired electrochemical reactions, electrocatalytic reactions, reactions undertaken in ionic liquids, electrogenerated reactants from renewable biomass, green organic electrosynthesis in micro- and nano-emulsions, and the synthesis of complex molecules with an electrosynthetic key step. Lastly, this chapter discusses potential future perspectives [25].

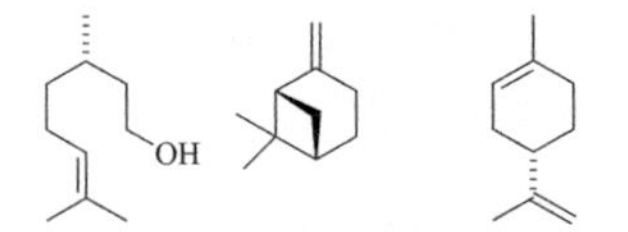

FIGURE 1.3 Chemical structure of (-)-citronellal, β-pinene, and (-)-limonene.

In recent years, natural terpenes have significantly influenced the field of chemistry, particularly in the organic synthesis of desirable molecules [26–31]. Whether in their linear form (as in the case of citronellol), circular form (as in limonene), or complex form (like β-pinene, shown in Figure 1.3), terpenes have been extensively employed in organic synthesis to develop various products, both natural and synthetic, with diverse applications [29, 32–36]. There is vast literature illustrating the significant potential of terpenes as building blocks for conventional organic synthesis of more intricate terpenes, as well as other types of molecules [29, 32–40].

Despite numerous improvements in synthesis techniques, the application of terpenes as raw materials for eco-friendly chemical processes is still being pursued and developed today. This chapter has solely concentrated on the use of terpenes as fundamental components in eco-friendly procedures, ensuring adherence to one or more principles of green chemistry [41].

1.5.1 Terpenes as Synthons for Green Chemistry

The profusion of functional groups present in terpenes, including double bonds, OH groups, and COOH groups, creates an opportunity to incorporate new functional groups, and this can lead to the generation of diverse and novel molecules. Nevertheless, the conversion of olefins through oxidative and reductive procedures may face difficulties due to potential conflicts with other reactions, including structural rearrangements, isomerization, and nucleophilic addition. De Oliveira and colleagues successfully transformed camphene 1 into epoxy camphene 2a, glycol camphene 2b, and aldehyde camphene 2c via a metal-free reoxidation technique, utilizing the $PdCl_2 / H_2O_2 / CH_3CN$ system as described in **Figure 1.4** [42]. The selectivity of oxidation products was found to decrease due to the water generated during the reaction, as well as the presence of water in the H_2O_2 solution. Utilizing the same system, the conversion of β-pinene 3 resulted in the formation of epoxy-β-pinene 2a, pinocarveol 2b, and pinocarvone 2c, with pinocarveol 2b being the predominant compound consistently observed. In contrast to camphene 1, which achieved a conversion rate of more than 90%, the conversion rate of β-pinene was restricted to 58% despite the usage of an excess amount of H_2O_2. The system did not yield any oxidation products upon the utilization of α-pinene and 3-carene. The existence of double bonds situated within the ring structure, as well as the presence of two sizeable methyl groups, may have impeded or even obstructed the oxidation process from happening [42].

Vieira and colleagues successfully accomplished the hydroformylation of terpenes aimed at obtaining natural aroma aldehydes, utilizing a biphasic system consisting of water and toluene [43]. The hydroformylation of β-citronellene (5) was carried out by

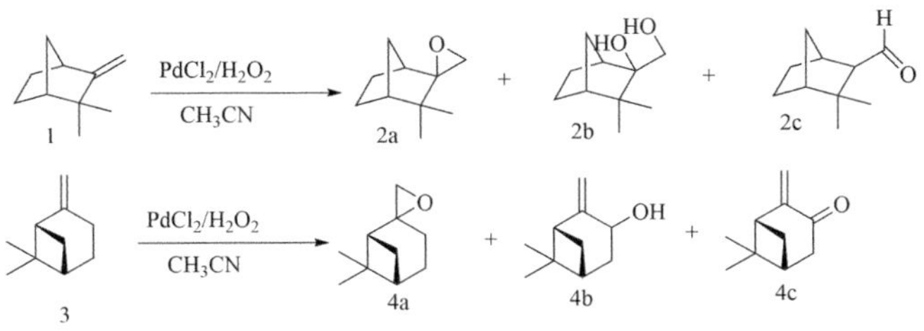

FIGURE 1.4 Oxidation of camphene (1) and β-pinene (3).

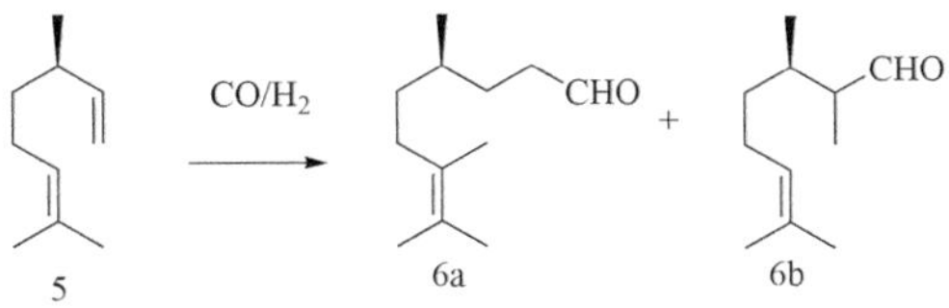

FIGURE 1.5 Hydroformylation of β-citronellene (5).

utilizing $[Rh(COD)(OMe)]^2$ as the catalyst precursor and a water-soluble tris(3-sulfonatophenyl)phosphine trisodium salt (TPPTS) as the ligand in a biphasic system composed of water and toluene, as depicted in **Figure 1.5**. During this eco-friendly hydroformylation process, the main product – the linear aldehyde 6a – in addition to the branched isomer 6b, was extracted from the organic phase [43]. The catalyst-containing aqueous layer can be separated and recycled for subsequent use [43].

When utilizing the same parameters, hemiacetals 8a and 10a were identified as the principal products for linalool 7 and nerolidol 9, respectively, as illustrated in Figure 1.6. The formation of the hemiacetals is a result of the self-cyclization of the respective hydroxyl-aldehydes. Himachalene isomers (11 and 12) originating from the ground wood of Atlantic cedar were employed as initial reagents to procure himachalene epoxides, which were later identified as minor elements found in essential oils [44]. As depicted in Figure 1.7, Loubidi and colleagues managed to extract himachalene epoxides by means of a solvent-free oxidation technique utilizing tert-butyl hydroperoxide as an oxidizing agent and [MoO2 [45]]2 as the catalyst [46]. Epoxides 13a and 13b were obtained in a 33/67 ratio [47].

Hydrotalcites were exploited as a catalyst, and hydrogen peroxide was utilized as an oxidizing agent during the Baeyer–Villiger oxidation of carvone 14 to extract compounds 15, 16, and 17 [48]. The yields achieved were moderate to low, as shown in Figure 1.8 by Rodilla and colleagues [49]. During the oxidation process of the separated epoxide 2 using hydrotalcites and meta-chloroperbenzoic acid, opening and lactonization took place, leading to a racemic blend of compounds 19 and 20.

An effective method for the orthoester Johnson–Claisen rearrangement has been established by Zlotin and colleagues [50] for the synthesis of naturally occurring terpenol derivatives, utilizing [bmim][PF6], an accessible ionic liquid, as a reusable catalyst. Linalool 7, isophytol 22b, myrtenol 24, and tetrahydronerolidol 22a yielded

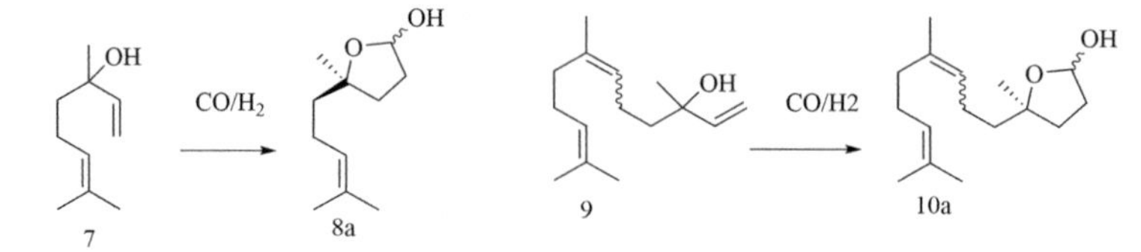

FIGURE 1.6 Linalool (7) and nerolidol (9) hydroformylation.

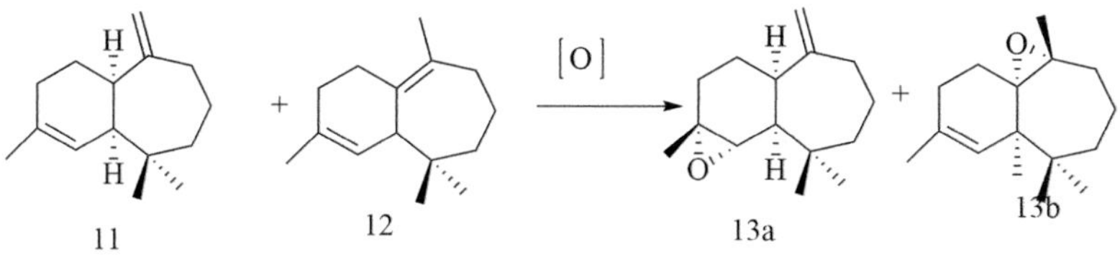

FIGURE 1.7 Himachalenes (11) and (12) epoxidation.

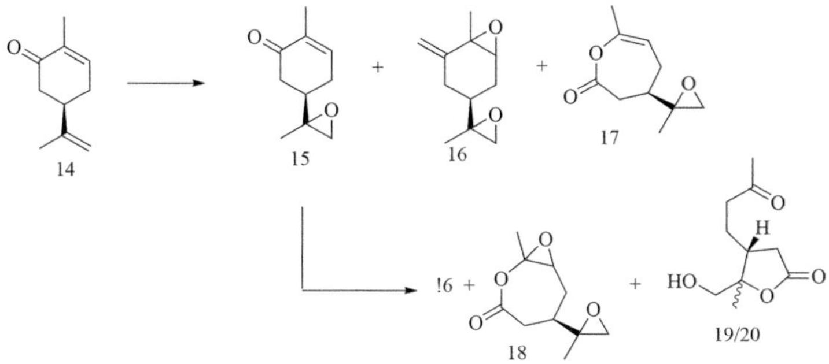

FIGURE 1.8 Carvone 14 oxidation.

rearrangement products (21, 23, 25) with satisfactory to excellent yields, as shown in Figure 1.9. Compound 21a, which comprises a geranylacetic acid structural fragment present in several bioactive substances, is produced via the reaction between linalool 7 and triethyl orthoacetate (where $R = H$), as illustrated in Figure 1.9 [51].

Numerous terpenic substances are manufactured using mineral acids as catalysts under homogeneous acid catalysis, which involves working with readily available precursors. Cotta and coworkers [52] have established a different method of heterogenous acid catalysis for β-pinene 3 and linalool 7 which involves using silica-supported tungstophosphoric heteropoly acid $H_3PW_{12}O_{40}$ (HPW) and benzaldehyde (where R = H) to generate oxabicyclo[3.3.1]nonene (as depicted in Figure 1.10).

Under identical conditions, using green solvents such as diethyl carbonate or dimethyl carbonate, a range of monoterpenic alcohols and monoterpenes reacted with cuminaldehyde (where R = Me) or benzaldehyde to yield the same oxabicyclo compound, as presented in Figure 1.10 [53].

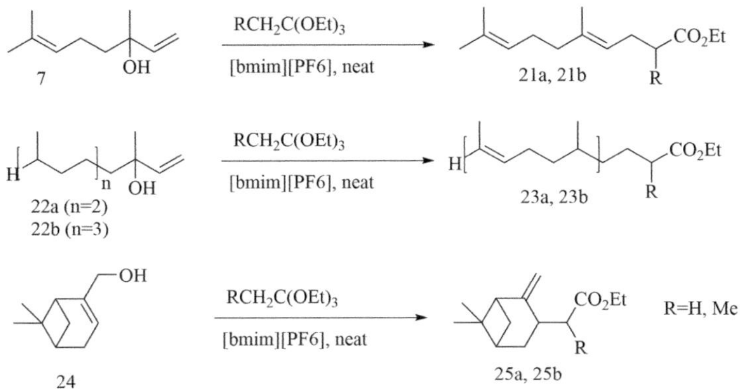

FIGURE 1.9 Johnson–Claisen rearrangements of linalool 7, tetrahydronerolidol 22a, isophytol 22b, and myrtenol 24.

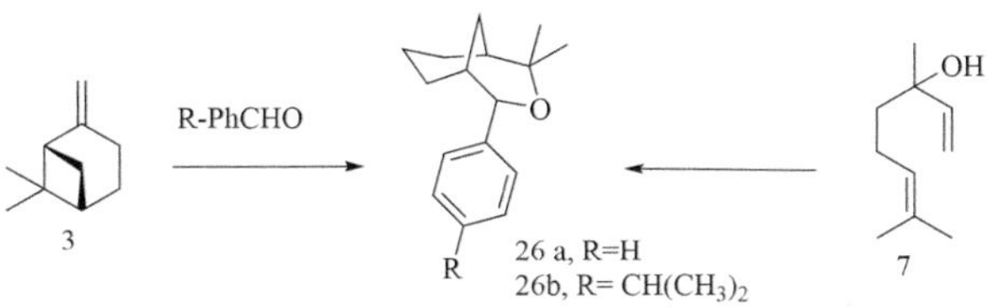

FIGURE 1.10 Cycloaddition of β-pinene (3) and linalool (7) by benzaldehyde and cuminaldehyde.

FIGURE 1.11 Hydrodeoxygenation of eucalyptol 27 via RuNPs@SILP.

Under identical conditions, using green solvents such as diethyl carbonate or dimethyl carbonate, a range of monoterpenic alcohols and monoterpenes reacted with cuminaldehyde (where R = Me) or benzaldehyde to yield the same oxabicyclo compound, as presented in Figure 1.10. Luska and colleagues [27] have created bifunctional catalysts using an acid-functionalized supported ionic liquid phase and ruthenium nanoparticles (RuNPs@SILP) to convert eucalyptol 27 via hydrodeoxygenation reaction into p-menthane 28 with remarkable efficiency (as shown in Figure 1.11). Hydrocarbons like menthane obtained from sustainable biomass sources can serve as an environmentally friendly substitute for fossil fuels. Certain naturally occurring hydrocarbons characterized by particular structures, such as pinanes, are highly coveted in various industries, including fine chemistry, pharmaceuticals, bioenergy, and food flavoring [28, 29]. Due to the inadequate presence of cis- and trans-pinane isomers in natural essential oils, the manufacturing sector has to resort to using artificially produced substitutes.

One of the primary hurdles in microbial electrosynthesis (MES) and other technologies that make use of CO_2 is the challenge of efficiently deriving high-value chemicals with multiple carbon atoms from it. In this study, we showcase a two-step process for generating terpenoids (>C20) and yeast biomass starting from CO_2 by integrating microbial electrosynthesis (MES) and yeast cultivation techniques. During the initial phase, *Clostridium ljungdahlii* is utilized to transform $CO_2(C_1)$ into acetate (C_2) through microbial electrosynthesis (MES). Subsequently, *Saccharomyces cerevisiae* is employed in the second stage to directly employ the acetate as a source of nourishment for generating sclareol (C_{20}), β-carotene (C_{40}), and yeast biomass. *S. cerevisiae* was able to manufacture 0.32 ± 0.04 mg/L of β-carotene, 2.54 ± 0.91 mg/L of sclareol, and 369.66 ± 41.67 mg/L of biomass utilizing the residual medium (1.5 g/L) containing unpurified acetate from MES reactors. According to the initial economic assessment, it appears that sclareol can be feasibly manufactured by employing recombinant *S. cerevisiae*, whereas non-recombinant *S. cerevisiae* can be utilized to generate biomass directly from the unpurified medium comprising acetate in the aftermath of MES [54].

Fenchone and menthone, which are cycloalkanones belonging to the terpenoid class, were subjected to oxidation using a platinum anode in an electrolyte environment that was either neutral or alkaline. The electrolyte $NaClO_4$ was used to produce 1-isopropyl-4-methyl-2-oxa-bicyclo[2.2.1]heptan-3-one from Fenchone (1a), and 6-isobutyl-4-methyl-tetrahydropyran-2-one from menthone (1b) in high yield. On the other hand, when Na_2CO_3 was used as electrolyte, 1a was oxidized to produce 2,2-dimethyl-3–(2-oxopropyl) cyclopentanone in a single step. The chapter includes suggestions for the reaction mechanisms [55].

In this section of the book, the most recent advancements in electrochemical techniques used to modify beta-lactams and terpenoids are discussed. An important factor in this process is to selectively modify the targeted heteroatom in intricate molecules while abstaining from altering other susceptible functional groups. To achieve this goal, a novel electrochemical approach has been created which involves generating reaction sites in electrolytic media, introducing new mediator systems, utilizing two-phase electrolysis, and other related techniques. Moreover, this text highlights the importance of employing reusable mediator substances and extremely efficient catalysts in the production of terpenoids, particularly during critical stages of the conversion process [56].

1.5.2 Conversion and Functionalization of Terpenoids

In recent times, there has been significant development in the domain of electro-organic chemistry, specifically in the realm of electrochemical conversions of biomolecules found in nature [57]. The current issue of concern pertains to understanding the ability of electro-organic reactions on particular groups of biomolecules, particularly in the terpenoid groups. The aim is to shed light on the involvement of electrochemistry in the synthesis of terpenoids.

1.5.2.1 Carbon–Carbon Bond Cleavages

The most preferred approach to produce oxoalkanoates and alkanediones, which serve as chiral building blocks for synthesizing different compounds, is through oxidative cleavage of carbon–carbon bonds in vicinally oxygenated cyclic terpenoids. The electrochemical cleavage reactions of carbon–carbon bonds have been studied on 1,2-diols [58, 59], 2-amino-l-ols [56, 60], and 1,2-diamines [60]. In recent times, electro-oxidative cleavage processes have been implemented on 2-oxo or 2-phenylthio-l-cycloalkanols [61], enol acetates [62] or enol ethers of cycloalkanone [63], and alpha-, beta-epoxycycloalkanones [64]. These processes result in the creation of the relevant oxoalkanoates or alkanediones [56]. The substance (+)-4-hydroxy-p-menth-3-one (29), when subjected to electrolysis in a system comprising $MeOH - LiClO_4$ -Pt at 20 V, produces methyl (-)-6-oxo-6,7-dihydrocitronellate (30) in a yield of 94% [62]. This process is depicted in Figure 1.12. The process can be effectively utilized in the synthesis of (+)-rose oxide 33. By subjecting 31 to electrolysis in an $MeOH - LiClO_4\text{-}(Pt)$ system at a temperature range of 8–10°C, it is possible to obtain the desired chiral intermediate 32 with an 84% yield according to the procedure outlined in Figure 1.13.

The process of electro-oxidative cleavage of alpha-, beta-epoxycycloatkanone systems has been employed in creating optically active methyl trans- and cis-chrysanthemates 32 starting from (+)- and (-)-carvones 33. The method is outlined in Figure 1.14 and is mentioned in reference [64]. At a low temperature range of 2–5°C and an anode potential of 2.10–2.23 V vs. $Ag / Ag+$, the double cleavage of 5–(1-chlorol-methylethyl)-2,3-epoxy-2,3-dimethylcyclohexan-l-one 34, occurring at the C-1/C-2 and C-2/C-3 bonds, can be achieved using an MeOH- or MeOH/AcOEt (7: 1)-$LiClO_4$-(Pt) system. The effectiveness of the current is notably increased by including a cosolvent that enhances the solubility of 34 in the solution.

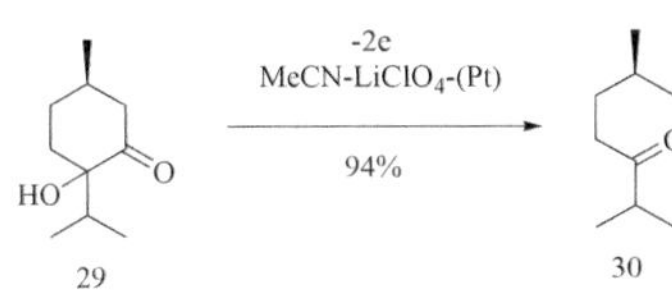

FIGURE 1.12 Electrochemical transformation of (+)-4-hydroxy-p-menth-3-one (29) in a MeOH–LiClO4–Pt system at 20 V, affording methyl (–)-6-oxo-6,7-dihydrocitronellate [30].

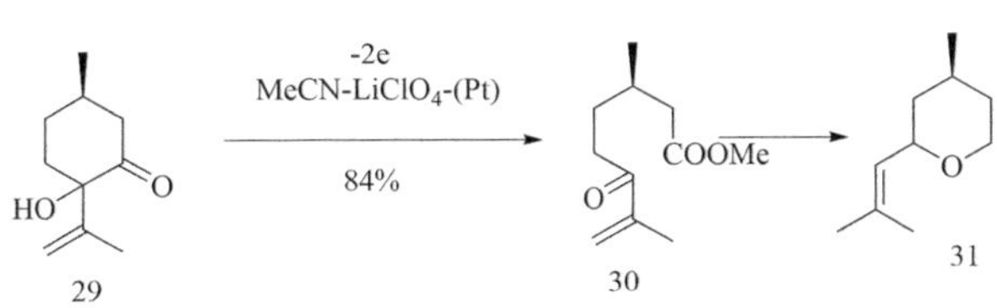

FIGURE 1.13 Electrolysis of 31 at 8–10 °C affords chiral intermediate 32 in 84% yield, enabling synthesis of (+)-rose oxide [33].

The process of anodic intramolecular coupling of l-benzyltetrahydroisoquinoline-type compounds to create morphinandienone alkaloids is not always a straightforward mechanism [111]. The mechanism of reactions that take place without the presence of acid cannot be accurately explained due to the indistinct nature of the cyclized products, resulting in ambiguity in the mechanism. The process carried out in acidic conditions has been thoroughly investigated. There are two potential ways for the beginning stage, either by oxidizing the aromatic ring system or by eliminating a nonbonding electron from the amino nitrogen atom. The likelihood of the second pathway is considerably low because the protonation hinders the nonbonding electrons. It has been confirmed through experimentation that the initial oxidation process does not involve the deprotonation of the nitrogen atom, as the intramolecular product obtained from deuterated O-benzyl-pseudo-codamine preserved over 95% of deuterium. Figure 1.22 illustrates the most likely path, consisting of three reversible steps, followed by the rate-determining electron transfer solution [112].

The anodic intramolecular coupling of 1-benzyltetrahydroisoquinolines results in flavinantine-type products through para-para coupling at the 2' position. However, coupling at the 6' position through para-ortho coupling is more preferable as it leads to products with a morphine-type alkaloid substitution pattern, shown in Figure 1.23.

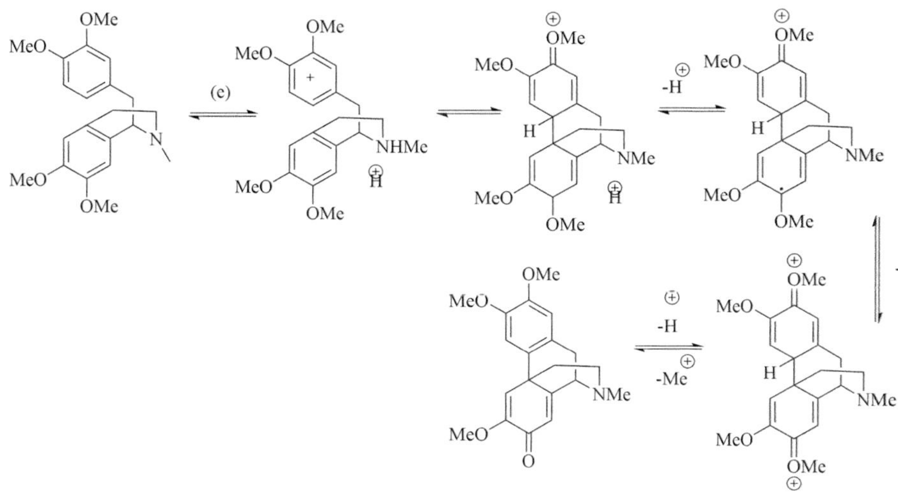

FIGURE 1.22 Path, consisting of three reversible steps.

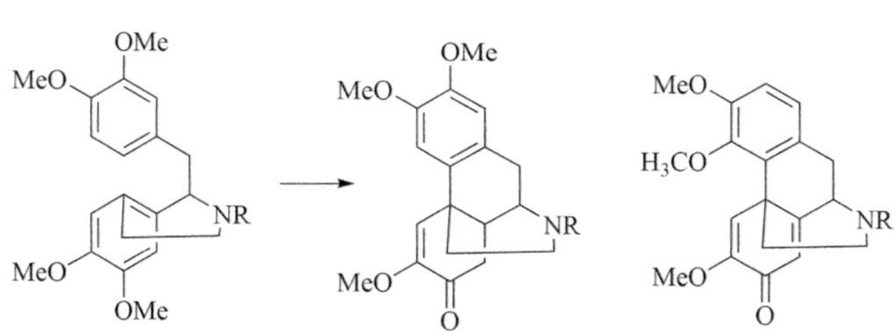

FIGURE 1.23 Flavinantine-type products.

Numerous modifications have been made to promote para-ortho coupling while preventing para-para coupling. Attempts to use halogen substituents on the 2' or 3' position of the benzyl ring as blocking groups were unsuccessful and resulted in the formation of cleavage or substitution products [113]. The use of 2'-nitro or 2'-N-acetylamino substituents resulted in the cleavage at C–1. A change in the reaction's outcome occurred when the amino group underwent trifluoroacetylation, resulting in the formation of neospirodienones instead of morphinandienones Figure 1.24. The modification in the reaction's outcome may be triggered by the reorganization of the intermediate cation, which is of the morphinandienone type. By introducing methanol to the reaction system as a nucleophile, it becomes feasible to capture the intermediate cation before it undergoes reorganization. Nevertheless, the para-para coupling pattern remains unchanged [114].

The initial substance used in the synthesis is a l-(3,5-dibenzyloxybenzyl)tetrahydroisoquinoline derivative, which fulfills three requirements, promoting intramolecular ortho-para coupling. Firstly, both benzene rings are activated equally by alkoxy groups. Secondly, the symmetrically substituted benzyl ring has identical coupling sites at 2' and 6'. Lastly, para-alkoxy groups activate both coupling positions. The desired para-ortho coupling product, 83, was obtained with a yield of 66% by subjecting the starting compound, 82, to anodic oxidation at 1.0 V in a mixture of CH_3OH and CH_3CN in a ratio of 4:1. The product 83 can be easily converted into salutaridinol 84 and 2-hydroxy-3-desoxythebaine 85 [66]. When 4-(3,4-dimethoxybenzyl)6,7-dimethoxy-2-methyl-l,2,3,4-tetrahydroisoquinoline is subjected to anodic oxidation, it does not produce aryl-aryl coupled products within the molecule; instead, it yields the salts of 3,4-dihydroisoquinolinium 86 and 4–(3,4-dimethoxybenzylidene)-6,7-dimethoxy-2-methyl-l,4-dihydroisoquinolinium 87, shown in Figure 1.25 [45].

Regarding the anodic intramolecular coupling of 1-benzyltetrahydroisoquinoline derivatives 88, there has been significant research on the anodic oxidation of 4-(3,4-dimethoxybenzyl)-6,7-dimethoxyisochromanone 89. The primary output derived from the isochromanone compound is a delta-lactone, known as 7a,8-dihydro-3,10,11-trimethoxy-2 H phenanthro [9.8a, b] furan-2,7(5 H)-dione 90 [66].

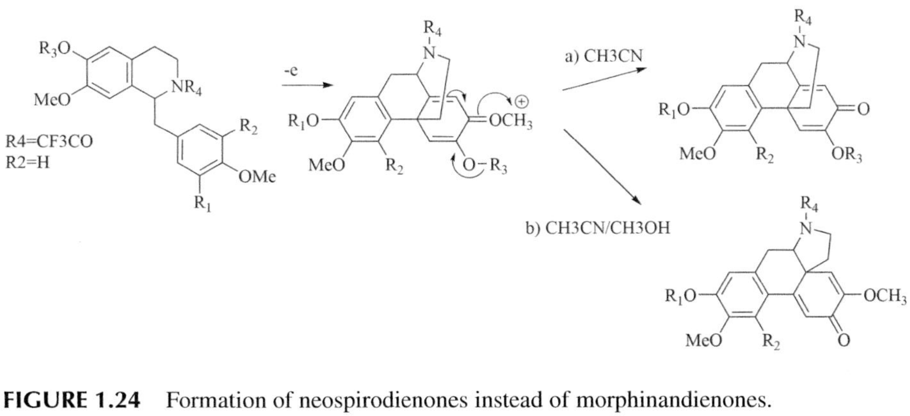

FIGURE 1.24 Formation of neospirodienones instead of morphinandienones.

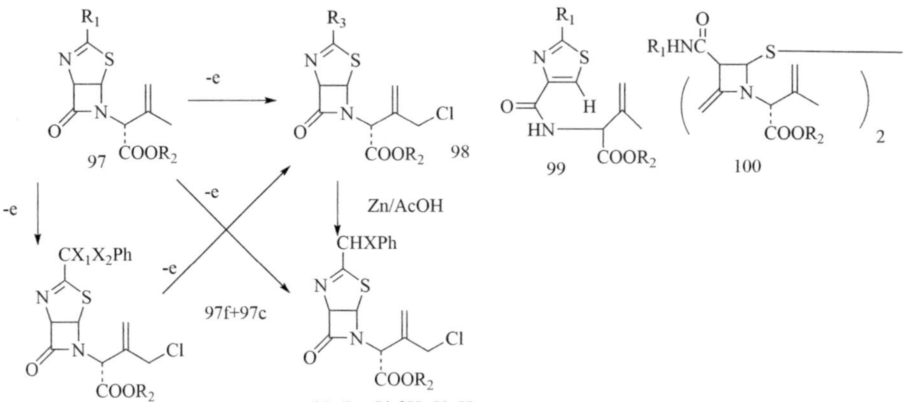

97c R_2=$PhCH_2$;X_1=X_2=Cl
97f R_2=$PhCH_2$; X_1=H;X_2=Cl
97g R_2=Me; X_1=X_2=Cl
97h R_2=X_1=H; X_2=Cl

98e R_2=$PhCH_2$;X=H
98f R_2=$PhCH_2$; X=Cl
98g R_2=Me; X=H

FIGURE 1.29 Thiazolin-azetidinones electrochlorination.

Instead of CH_2Cl_2, other types of hydrophobe solvents, such as $CHCl_3$ and AcOEt, could also be utilized. If hydrophilic solvents such as THF, CH_3CN, CH_3CN/THF, or CH_2Cl_2 / THF are used, even in a dual-phase system, it could help in the hydrolysis of the thiazoline or beta-lactam ring, leading to the developed products, 99 and/or 100, that have an open ring. During the electrochlorination process of 97b (R_1 = $PhCH_2$), when a larger volume of sodium chloride (1 g/3 mL) is used in water (entry 3), the production of gem-dichloride lc (89%) as an initial product is observed when 10 F/mol of electricity is passed (as shown in Figure 1.29). The process of electrolysis of the dichloride lc in the same substance resulted in the creation of 2b, giving a yield of 85% (demonstrated in entry 4). The difference in the distribution of products is believed to be due to the various chlorination agents created at the anode by the discharge of Cl-, such as Cl_2, HOCl, Cl_2O, etc. This factor is dependent on the concentration of Cl-, the pH value of the substance, the oxidation potentials, and the kind of aprotic solvents present [122]. The illustration in Figure 2.1 indicates that when chloride ion is discharged at the anode, it forms chlorine molecules. These molecules can be hydrolyzed in an aqueous layer to produce hypochloric acid if the substance maintains a weakly acidic pH value and a chloride ion concentration that is lower than 100 mg/3 mL. It is a recognized fact that hypochloric acid is in a state of equilibrium with chlorine oxide (Cl_2O). Subsequently, chlorine as well as chlorine oxide would move toward the organic layer and react with olefins to create chlorination products, also referred to as ene-type products. The process of converting 98a and 98b $(R_3 = PhCCl_2)$ into their corresponding allylic chlorides, 2e and 98g $(R_3 = PhCH_2)$, is successful, with a yield of over 90%. This is done through the elimination of benzyl chlorine atoms by utilizing zinc dust reduction in an AcOH / CH_2Cl_2 (1/4) system [121].

The thiazoline-azetidinones 98 can be transformed directly into 3'-thio-substituted cephalosporin 102. This is achieved through the opening of the thiazoline portion with sulfenyl chloride, then closing the ring with ammonia in dimethylformamide while also replacing the allylic chlorine atom with leaving thiolates [123]. When thiazoline-azetidinone homolog 97 is subjected to photo-induced electrochlorination, the resulting product is typically pentachloride 103. However, in an $HzO/CHC1_3-NaCI/H_2SO_4\text{-}(Pt)$ two-layer system, the initial chlorination reaction leads to the formation of trichloride 98 (with $R_3 = PhCCl_2$) in a yield of 80% after 15 F/mol of electricity is passed through the system. If the electrolysis process, which is carried out under the irradiation of a 750 W halogen lamp, is prolonged without isolating the intermediates, it results in an overall yield of 50% of pentachloride 103. The transformation of 97 into 103 can also occur when subjected to light in the presence of chlorine, while being dissolved in chloroform. This method yields 90% of pentachloride 103. C-2 substituted cephalosporins can be synthesized using compound 103 [123].

1.6.2.2 Electroreductive Removal of C-3 and C-3 Substituents of Cephalosporins

1.6.2.2.1 Reductive Removal of C-3′ Substituents of Cephalosporins (Synthesis of 3-Methylenecephams)

3-methylenecepham 105 is beneficial in creating 3-norcephalosporins, a group of beta-lactam antibiotics that involve heteroatoms, such as chlorine or methoxy, in the C-3 position. For the preparation of 3-methylenecepham 105 (in Figure 1.30), 3'-acetoxycephalosporin 104 serves as a suitable precursor.

In 1972, Ochiai and their colleagues conducted the electrochemical reduction method to create 3-methylenecepham derivatives [56]. When 3'-acetoxycephalosporanic acid 104 is subjected to electrolysis using aqueous $HCl\text{-}Na_2HPO_4$ (0.1 M, pH 7) and a Pt/Hg catalyst, it results in a yield of 81% of the corresponding 3-methylenecepham 105. The process of electrochemical reduction for cephalosporin derivatives has been researched separately to understand its mechanism [124].

An alternate simple method to obtain 105 is through the electrochemical reduction process that removes chlorine from 3'-chlrocephalosporin 106 [125]. In a divided cell system using a THF/H20 (4/1)-$LiClO_4/NH_4ClO_4\text{-}(Pt/Pb)$ catalyst, the electrolytic conversion of 106 produces a yield of 87% of 105. Choosing the correct cathode material is crucial for the process, and it has been observed that lead cathode is necessary

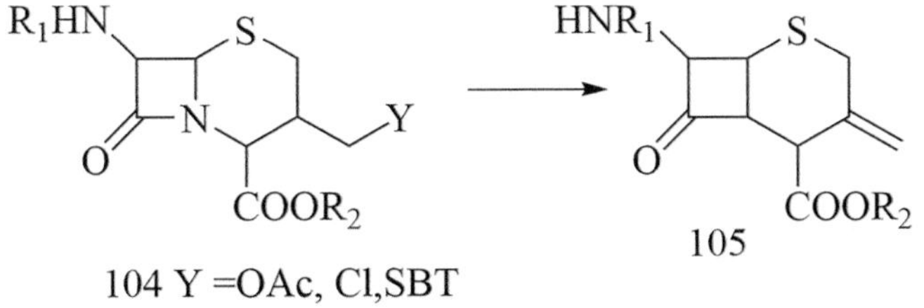

FIGURE 1.30 3-methylenecepham 105, 3'-acetoxycephalosporin 104.

for the effective conversion of 104 to 105, as the yield of 105 changes depending on the choice of cathode materials. In particular, the order of effectiveness of cathode materials is as follows: Pt (19%), Cu (51%), C (58%), and Pb (87%). Likewise, the conversion of Y-(2-benzothiazolylthio)-cephalosporin 104 (Ar:Benzothiazolyl) also results in 78–90% yields of 3-methylenecepham 105 using a similar process. The creation of 3-methylenecepham 105 has also been accomplished through the electrochemical reduction method of cephalosporin S-oxide 106 and 107, as shown in Figure 1.31 [56].

The field of electrochemical conversion of naturally occurring biomolecules has observed significant advancements in electro-organic chemistry recently [56]. The current focus is on exploring the ability of electro-organic reactions to synthesize certain biomolecules, especially in the field of terpenoids. This explanation aims to clarify the involvement of electrochemistry in the terpenoid area.

Multiple effective approaches exist for electrochemically reducing prochiral substrates, thereby creating optically active end products [126]. The technique most commonly adopted for this task is to reduce the substrate in the presence of a chiral alkaloid acting as an inductor at a mercury pool, as stated in references [127–131]. When the alkaloid is not firmly adsorbed at the mercury cathode, it is utilized in large quantities and acts as the supporting electrolyte. In scenarios where the alkali is firmly adsorbed, only a slight amount is required to achieve maximum optical yield.

We have demonstrated recently that by utilizing a solution of 10^{-4} mol/L strychnine during the electroreduction of both phenylglyoxylic acid and the corresponding oximes (anti- and syn-isomers), optically active mandelic acid and C-phenylglycine, respectively, can be produced with a maximum optical yield of approximately 20%. In addition, we have verified that the favored configuration of the reduced products and maximal optical yield are dependent on the pH level of the solution and the cathodic potential [129, 130, 132]. The statement put forward is that if the quantity of the inductor is raised, it can cause the electroreduction's stereochemical pathway to become reversed.

Other situations involving asymmetric electrosynthesis have been examined to determine the impact of alkaloid concentration on *p*. While a minor reduction in *p* was seen in these cases, there was no mention of a change in stereochemistry. Our findings demonstrate that the favored arrangement of the result is not solely determined by how the alkaloid and the substrate's structures interact with each other.

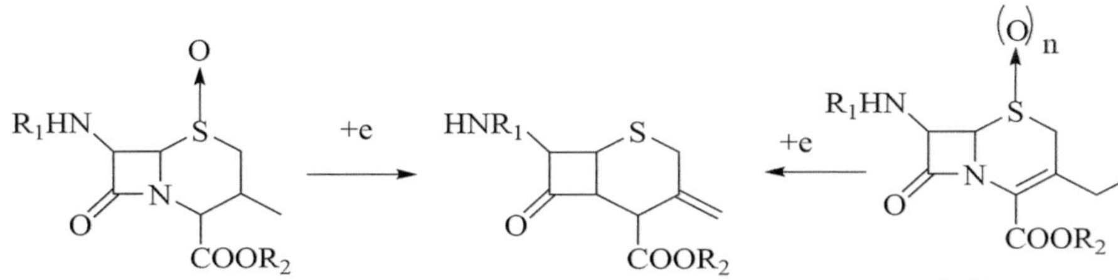

106 105 107

FIGURE 1.31 Conversion and functionalization of terpenoids.

Other aspects, such as the potential for the intermediate carbanion to invert within the electric field of the electrode before protonation, may significantly affect the outcome. The chiral alkaloid acting as an acid cannot be the sole explanation for the intermediate carbanion's protonation, which does not account for the inversion of configuration when increasing the alkaloid concentration or making the potential more negative at the electrode surface. Experimental data on the behavior of the alkaloid is still needed to fully understand its effect on stereochemistry, similar to the case of cathodic hydrogenation of carbon–carbon double bonds, where protonation by a protonated alkaloid was not found to be the determining step [126].

There has been a lot of attention given to the field of asymmetric electrosynthesis recently. One of the methods that has been explored is the reduction of prochiral ketones or oximes to produce optically active alcohols or amines with the aid of a small amount of adsorbed alkaloid during the electrochemical process [133]. This approach has been heavily researched by Jubault et al. [129]. The electrochemical reduction of phenylglyoxylic acid and its oxime counterparts (both anti- and syn-isomeric forms), conducted in the presence of strychnine at a concentration of 10^{-4} mol/L[1], affords enantioenriched mandelic acid and C-phenylglycine, respectively, with an optical yield (p) approaching 20%. Moreover, it was established that both the predominant stereochemical configuration of the reduction products and the magnitude of p are critically contingent upon the solution's pH and the applied cathodic potential. Further disclose that modulation of the inductor concentration can induce an inversion in the stereochemical trajectory of the electroreduction. Experimental validation was undertaken with the two substrates previously examined – phenylglyoxylic acid and the anti-oxime derivative. Controlled-potential electrolyses were executed at a mercury pool electrode within buffered acetic acid media in the presence of strychnine, and the dependence of p on inductor concentration was systematically delineated across varying cathodic potentials [126].

According to the table of results on this chapter, the process of asymmetrical induction takes place in solutions with mild acidic or basic properties. The reason for the asymmetric induction could be due to the selective protonation of the two enantiotopic faces of the oxime double bond, which is governed by the chiral acid properties of the protonated alkaloid [126]. The chemical reaction produces an 'adduct' between the substrate and the inductor, which remains in this state until the intermediate carbanion is finally protonated by a protonating species located at the electrode surface [126]. This preliminary step of asymmetric induction, which causes one of the enantiotopic faces to be presented more prominently, is similar to Horner's proposal. In this study, we have presented additional evidence to support the mechanism behind the induction of asymmetry. Specifically, we propose that the preprotonation process may strengthen the preferential orientation of the substrate concerning the adsorbed inductor [126].

1.7 CONCLUSION

To demonstrate the value of biomolecules which possess a natural origin or are made by fermentation procedures is one of the attractive topics in the field of organic synthesis. In recent decades, regarding organic synthesis, the electrochemical methods

7-dimethoxy-2-methyl-1, 2, 3, 4-tetrahydroisoquinoline and attempted synthesis of 4–(3, 4-dimethoxybenzyl) 6, 7-dimethoxy-2-methyl-1, 4-dihydro-3 (2 H)-isoquinolone. *Journal of the Chemical Society, Perkin Transactions*. 1980: p. 2013–2020.
46. Loubidi, M., et al., Solvent-free epoxidation of himachalenes and their derivatives by TBHP using [MoO2 (SAP)] 2 as a catalyst. *Comptes Rendus. Chimie*. 2014. **17**(6): p. 549–556.
47. Li, Y. and F. Chemat, *Plant based "green chemistry 2.0"*. 2019. Singapore: Springer.
48. Rives, V., *Layered double hydroxides: present and future*. 2001. Nova Publishers.
49. Rodilla, J.M., et al., Hydrotalcite catalysis for the synthesis of new chiral building blocks. *Natural Product Research*. 2016. **30**(7): p. 834–840.
50. Kryshtal, G.V., et al., The orthoester Johnson-Claisen rearrangement of allylic terpenols in the presence of acidic ionic liquid. *Journal of Fluorine Chemistry*. 2016. **183**: p. 23–29.
51. Zlotin, S.G., et al., Kinetic resolution of racemic (cyclohexyl)(geranyl) acetic acid. *Mendeleev Communications*. 2014. **5**(24): p. 257–259.
52. Cotta, R.F., et al., Coupling of monoterpenic alkenes and alcohols with benzaldehyde catalyzed by silica-supported tungstophosphoric heteropoly acid. *Catalysis Today*. 2017. **289**: p. 14–19.
53. Cotta, R.F., et al., Heteropoly acid catalysts in upgrading of biorenewables: cycloaddition of aldehydes to monoterpenes in green solvents. *Applied Catalysis B: Environmental*. 2017. **217**: p. 92–99.
54. Yadav, R., et al., Microbial electrosynthesis from carbon dioxide feedstock linked to yeast growth for the production of high-value isoprenoids. *Bioresource Technology*. 2022. **363**: p. 127906.
55. Batanero, B., J. Recio, and F. Barba, One-pot anodic lactonization of fenchone and menthone and electrosynthesis of a new magnolione analogue. *Electrochemistry Communications*. 2016. **66**: p. 29–33.
56. Torii, S., H. Tanaka, and T. Inokuchi, Role of the electrochemical method in the transformation of beta-lactam antibiotics and terpenoids. *Electrochemistry III*. 1988: p. 153–194.
57. Lévêque, J.-M., et al., *Organic sonochemistry: challenges and perspectives for the 21st century*. 2018. Springer.
58. Shono, T., Y. Matsumura, T. Hashimoto, K. Hibino, H. Hamaguchi and T. Aoki, Electroorganic chemistry. XXII. Novel anodic cleavage of glycols to carbonyl compounds. *Journal of the American Chemical Society*. 1975. **2546**.
59. Torii, S., et al., Efficient conversion of olefins into epoxides, bromohydrins, and dibromides with sodium bromide in water-organic solvent electrolysis systems. *The Journal of Organic Chemistry*. 1981. **46**(16): p. 3312–3315.
60. Portis, L.C., J.T. Klug, and C.K. Mann, Electrochemical oxidation of some phenethylamines. *The Journal of Organic Chemistry*. 1974. **39**(24): p. 3488–3494.
61. Torii, S., T. Inokuchi, and Y. Araki, Electrooxidative cleavage of carbon-carbon bonds. III. Preparation of alkanediones from cycloalkanones through 2-phenylthio-1-cycloalkanols. *Synthetic Communications*. 1987. **17**(15): p. 1797–1805.
62. Torii, S., T. Inokuchi, and R. Oi, Electrooxidative cleavage of carbon-carbon linkages. 1. Preparation of acyclic oxoalkanoates from 2-hydroxy-and 2-acetoxy-1-cycloalkanones and cycloalkanone enol acetates. *The Journal of Organic Chemistry*. 1982. **47**(1): p. 47–52.
63. Torii, S., et al., Anodic ring opening reaction of 2-methoxy-1-cyclopentene-1-carboxylic acids. *Bulletin of the Chemical Society of Japan*. 1973. **46**(8): p. 2534–2537.
64. Torii, S., T. Inokuchi, and R. Oi, Electrooxidative cleavage of carbon-carbon bonds. 2. Double cleavage of. alpha.,. beta.-epoxy alkanones and enantiospecific syntheses of chiral methyl trans-and cis-chrysanthemates from (+)-and (-)-carvones. *The Journal of Organic Chemistry*. 1983. **48**(12): p. 1944–1951.

65. Shono, T. Synthesis of alkaloidal compounds using an electrochemical reaction as a key step. In *Electrochemistry III*. 1988. Berlin, Heidelberg: Springer.
66. Rabaey, K. and R.A. Rozendal, Microbial electrosynthesis – revisiting the electrical route for microbial production. *Nature Reviews Microbiology*. 2010. **8**(10): p. 706–716.
67. Marshall, C.W., et al., Metabolic reconstruction and modeling microbial electrosynthesis. *Scientific Reports*. 2017. **7**(1): p. 1–12.
68. Rozendal, R.A., et al., Hydrogen production with a microbial biocathode. *Environmental Science & Technology*. 2008. **42**(2): p. 629–634.
69. Cheng, S., et al., Direct biological conversion of electrical current into methane by electromethanogenesis. *Environmental Science & Technology*. 2009. **43**(10): p. 3953–3958.
70. Nevin, K.P., et al., Microbial electrosynthesis: feeding microbes electricity to convert carbon dioxide and water to multicarbon extracellular organic compounds. *MBio*. 2010. **1**(2): p. e00103–e00110.
71. Steinbusch, K.J., et al., Bioelectrochemical ethanol production through mediated acetate reduction by mixed cultures. *Environmental Science & Technology*. 2010. **44**(1): p. 513–517.
72. Ishii, S.I., et al., A novel metatranscriptomic approach to identify gene expression dynamics during extracellular electron transfer. *Nature Communications*. 2013. **4**(1): p. 1–10.
73. Ishii, S.I., et al., Microbial metabolic networks in a complex electrogenic biofilm recovered from a stimulus-induced metatranscriptomics approach. *Scientific Reports*. 2015. **5**(1): p. 1–14.
74. Kiely, P.D., J.M. Regan, and B.E. Logan, The electric picnic: synergistic requirements for exoelectrogenic microbial communities. *Current Opinion in Biotechnology*. 2011. **22**(3): p. 378–385.
75. Tremblay, P.-L. and T. Zhang, Electrifying microbes for the production of chemicals. *Frontiers in Microbiology*. 2015. **6**: p. 201.
76. Kodama, S., et al., Synthesis of Amaryllidaceae alkaloids, siculine, oxocrinine, epicrinine, and buflavine. Tetrahedron. 2004. **60**(22): p. 4901–4907.
77. Proskurnina, N. and A. Yakovleva, About alkaloids Galanthus woronowii II. Allocating a new alkaloid. Zhurnal obshchej khimii. 1952. **10**(22): p. 1899.
78. Marco-Contelles, J., et al., Synthesis and pharmacology of galantamine. Chemical reviews. 2006. **106**(1): p. 116–133.
79. Gauthier, S., et al., journey through the diagnosis of dementia. *World Alzheimer Report*. 2021: p. 17–29.
80. Thomsen, T. and H. Kewitz, Selective inhibition of human acetylcholinesterase by galanthamine in vitro and in vivo. *Life Sciences*. 1990. **46**(21): p. 1553–1558.
81. Cimmino, A., et al., Amaryllidaceae alkaloids: absolute configuration and biological activity. Chirality. 2017. **29**(9): p. 486–499.
82. Rinner, U., C. Dank, and T. Hudlicky, Galanthamine. *Targets in Heterocyclic Systems*. 2016. **20**: p. 283–315.
83. Trost, B.M., W. Tang, and F.D. Toste, Divergent enantioselective synthesis of (-)-galanthamine and (-)-morphine. *Journal of the American Chemical Society*. 2005. **127**(42): p. 14785–14803.
84. Node, M., et al., Biomimetic synthesis of (+/-)-galanthamine and asymmetric synthesis of (-)-galanthamine using remote asymmetric induction. *Chemical and Pharmaceutical Bulletin*. 2006. **54**(12): p. 1662–1679.
85. Kita, Y., et al., Oxidative intramolecular phenolic coupling reaction induced by a hypervalent iodine (III) reagent: leading to galanthamine-type Amaryllidaceae alkaloids. *The Journal of Organic Chemistry*. 1998. **63**(19): p. 6625–6633.
86. Barton, D. and G. Kirby, 153. Phenol oxidation and biosynthesis. Part V. The synthesis of galanthamine. *The Journal of Organic Chemistry*. 1962: p. 806–817.

87. Kodama, S., et al., Total synthesis of (-)-galanthamine by remote asymmetric induction. *Angewandte Chemie International Edition*. 2004. **43**(20): p. 2659–2661.
88. Küenburg, B., et al., Development of a pilot scale process for the anti-Alzheimer drug (−)-galanthamine using large-scale phenolic oxidative coupling and crystallisation-induced chiral conversion. *Organic Process Research & Development*. 1999. **3**(6): p. 425–431.
89. Xiong, Z., et al., Biomimetic total synthesis of (−)-galanthamine via intramolecular anodic aryl-phenol coupling. *Organic & Biomolecular Chemistry*. 2022. **20**(20): p. 4123–4127.
90. Abernethy, D.R., et al., Metal impurities in food and drugs. *Pharmaceutical Research*. 2010. **27**(5): p. 750–755.
91. Beil, S.B., D. Pollok, and S.R. Waldvogel, Reproducibility in electroorganic synthesis-myths and misunderstandings. *Angewandte Chemie International Edition*. 2021. **60**(27): p. 14750–14759.
92. Pollok, D. and S.R. Waldvogel, Electro-organic synthesis – a 21(st) century technique. *Chemical Science*. 2020. **11**(46): p. 12386–12400.
93. Little, R.D. and K.D. Moeller, Introduction: electrochemistry: technology, synthesis, energy, and materials. *Chemical Reviews*. 2018. **118**(9): p. 4483–4484.
94. Mihelcic, J. and K.D. Moeller, Oxidative cyclizations: the asymmetric synthesis of (-)-alliacol A. *Journal of the American Chemical Society*. 2004. **126**(29): p. 9106–9111.
95. Rosen, B.R., et al., Total synthesis of dixiamycin B by electrochemical oxidation. *Journal of the American Chemical Society*. 2014. **136**(15): p. 5571–5574.
96. Pollok, D., et al., Allocolchicines – synthesis with electro-organic key transformations. *Organic Letters*. 2022. **24**(21): p. 3760–3765.
97. Lipp, A., et al., Total synthesis of (-)-oxycodone via anodic aryl-aryl coupling. *Organic Letters*. 2019. **21**(6): p. 1828–1831.
98. Lipp, A., et al., A regio- and diastereoselective anodic aryl-aryl coupling in the biomimetic total synthesis of (-)-thebaine. *Angewandte Chemie International Edition*. 2018. **57**(34): p. 11055–11059.
99. Seidler, J., et al., Does electrifying organic synthesis pay off? The energy efficiency of electro-organic conversions. *MRS Energy & Sustainability*. 2020. **7**: p. E42.
100. Pollok, D., et al., A general electro-synthesis approach to amaryllidaceae alkaloids. *Chemistry–A European Journal*. 2022. **28**(50): p. e202201523.
101. Barjau, J., G. Schnakenburg, and S.R. Waldvogel, Diversity-oriented synthesis of polycyclic scaffolds by modification of an anodic product derived from 2,4-dimethylphenol. *Angewandte Chemie International Edition*. 2011. **50**(6): p. 1415–1419.
102. Waldvogel, S.R., et al., Electrochemical arylation reaction. *Chemical Reviews*. 2018. **118**(14): p. 6706–6765.
103. Groß, J., J. Kühlborn, and T. Opatz, Applications of xylochemistry from laboratory to industrial scale. *Green Chemistry*. 2020. **22**(14): p. 4411–4425.
104. Bobbitt, J.M., et al., Electrochemistry of natural products. III. Stereoselective, stereospecific phenol coupling reaction. *Journal of the American Chemical Society*. 1971. **93**(14): p. 3551–3552.
105. Bobbitt, J.M., et al., Electrochemistry of natural products. II. Electrolytic oxidation of some simple 1, 2, 3, 4-tetrahydroisoquinoline phenols. *The Journal of Organic Chemistry*. 1971. **36**(20): p. 3006–3010.
106. Bobbitt, J.M., et al., Electrochemistry of natural products. V. Intramolecular coupling of phenolic alkaloid precursors. *The Journal of Organic Chemistry*. 1975. **40**(20): p. 2924–2928.
107. Miller, L., F. Stermitz, and J. Falck, Electrooxidative cyclization of 1-benzyltetrahydroisoquinolines: novel nonphenol coupling reaction. *Journal of the American Chemical Society*. 1973. **95**(8): p. 2651–2656.

108. Kotani, E., N. Takeuchi, and S. Tobinaga, Total synthesis of the alkloids (±)-oxocrinine and (±)-oxomaritidine via anodic oxidation. *Journal of the Chemical Society, Chemical Communications*. 1973. **15**: p. 550–551.
109. Kotani, E. and S. Tobinaga, Total synthesis of the morphinandienone alkaloids,(±)-amurine,(±)-flavinantine, and (±)-pallidine by anodic oxidation. *Tetrahedron Letters*. 1973. **14**(48): p. 4759–4762.
110. Kametani, T., K. Shishido, and S. Takano, Studies on the syntheses of heterocyclic compounds. Part DXCVII. A synthesis of dienone analogs by electrolytic oxidation. *Journal of Heterocyclic Chemistry*. 1975. **12**(2): p. 305–307.
111. Kotani, E., M. Kitazawa, and S. Tobinaga, A new synthesis of the alkaloid (±)-cryptopleurine via anodic oxidation. *Tetrahedron*. 1974. **30**(17): p. 3027–3030.
112. Kerr, J.B., T.C. Jempty, and L. Miller, Mechanistic study of anodic intramolecular coupling reactions. *Journal of the American Chemical Society*. 1979. **101**(24): p. 7338–7346.
113. Miller, L.L., et al., Synthesis of morphinandienones, a dihydrophenathrone, and Pummerer's ketones by anodic coupling. *The Journal of Organic Chemistry*. 1978. **43**(8): p. 1580–1586.
114. Klünenberg, H., C. Schäffer, and H.-J. Schäfer, A remarkable influence of the electrolyte in anodic cyclization of 1-benzyltetrahydroisoquinolins to neospirodienones or morphinandienones. *Tetrahedron Letters*. 1982. **23**(44): p. 4581–4584.
115. Christensen, L. and L.L. Miller, Electrochemical oxidation of morphinandienones. *The Journal of Organic Chemistry*. 1981. **46**(24): p. 4876–4880.
116. Shono, T., et al., Electrochemical oxidation of N, N-dimethylanilines. *Chemistry and Pharmacology of Naturally Occurring Bioactive Compounds*. 1982. **104**: p. 5753–5756.
117. Shono, T. and T. Shono, Anodic oxidations. In *Electroorganic chemistry as a new tool in organic synthesis*. 1984. Berlin, Heidelberg: Springer Berlin Heidelberg: p. 5–124.
118. Torii, S., *Electroorganic syntheses: oxidations*. Vol. 15. 1985. Wiley-VCH.
119. Torii, S., et al., Prominent aspects of electroorganic synthesis in β-lactam chemistry. *Bulletin des Sociétés Chimiques Belges*. 1982. **91**(12): p. 951–965.
120. Cooper, R.D. and F. José, Structural studies on penicillin derivatives. IV. Novel rearrangement of penicillin V sulfoxide. *Journal of the American Chemical Society*. 1970. **92**(8): p. 2575–2576.
121. Torii, S., et al., Chemoselective electrolytic chlorination of methyl group of 3-methyl-3-butenoate moiety of thiazoline-azetidinone homologues. *Tetrahedron Letters*. 1981. **22**(33): p. 3193–3196.
122. Muylder, J.V. and M. Pourbaix, *Atlas of electrochemical equilibria in aqueous solution*. 1966. Pergamon Press and Cebelcor: p. 287.
123. Torii, S., et al., Penicillin-cephalosporin conversion IV. Direct synthesis of 3'-thiosubstituted cephalosporins from thiazoline-azetidinones. *Tetrahedron Letters*. 1982. **23**(24): p. 2495–2498.
124. Hall, D.A., et al., The electrochemistry of cephalosporin C derivatives: part II. Cephalothin, sodium salt. *Journal of Electroanalytical Chemistry*. 1977. **80**(1): p. 155–170.
125. Torii, S., et al., Penicillin-cephalosporin conversion. XII. Electroreductive dechlorination of 3′-chlorocephalosporins into 3-methylenecephams. *Bulletin of the Chemical Society of Japan*. 1986. **59**(12): p. 3975–3976.
126. Jubault, M.J, Effect of alkaloid concentration in asymmetric electrosynthesis. *Journal of the Chemical Society, Chemical Communications*. 1980. **20**: p. 953–954.
127. Horner, L. and D. Degner, Studien zum vorgang der wasserstoffübertragung 26 Asymmetrische elektrochemische reduktion einiger alkyl-aryl-ketone über den einfluß der liganden im substrat-und leitsalzmolekül auf größe und richtung der asymmetrischen induktion. *Tetrahedron Letters*. 1971. **12**(17): p. 1245–1248.
128. Horner, L. and D.J.E.A. Degner, Studien zum vorgang der wasserstoffübertragung – 33. Zur kenntnis des aufbaus der elektrochemischen doppelschicht. Die asymmetrische

elektrochemische reduktion einiger alkyl-phenyl-ketone in gegenwart optisch aktiver leitsalze. *Electrochimica Acta*. 1974. **19**(10): p. 611–627.
129. Jubault, M., E. Raoult, and D. Peltier, Preprotonation of the substrate by protonated alkaloid in asymmetric electrosynthesis. *Journal of the Chemical Society, Chemical Communications*. 1979. (5): p. 232–233.
130. Kopilov, J., E. Kariv, and L.L. Miller, Asymmetric, cathodic reduction of acetylpyridines. *Journal of the American Chemical Society*. 1977. **99**(10): p. 3450–3454.
131. Hazard, R., S. Jaouannet, and A. Tallec, Synthese asymetrique par voie electrochimique: réduction du dibromo-1, 1 diphényl-2, 2 cyclopropane. *Journal of the American Chemical Society*. 1979. **20**(13): p. 1105–1108.
132. Jubault, M., et al., Effect of cathodic potential on the electrochemical synthesis of optically active amino-acids. *Journal of the Chemical Society, Chemical Communications*. 1977. **8**: p. 250–251.
133. Jubault, M., E. Raoult, and D.J.E.A. Peltier, Synthese asymetrique par voie electrochimique – III. stereochimie de la reduction du phenylglyoxylate de (−) menthyle. *Electrochimica Acta*. 1977. **22**(1): p. 67–73.

2 Biosynthesis of Nanoparticles and Applications in Electrochemistry

Mohammad Hadi Karimi Harandi[1]
[1]Department of Chemistry, Imam Hossein University, Tehran, Iran.

2.1 INTRODUCTION

2.1.1 Background and Significance of Nanoparticles in Electrochemistry

The introduction of nanoparticles into the field of electrochemistry was made possible by advances in microscopic tools and surface analysis techniques in the late 20th century. The development of atomic force microscopy (AFM) and scanning tunneling microscopy (STM) enabled the direct study of the structure and behavior of nanoparticles at the atomic scale. Nanoparticles, as one of the most important achievements of nanotechnology, have gained a key position in modern science and technology, especially in the field of electrochemistry, over the past two decades. The unique properties of these materials, resulting from their very small dimensions and very high surface-to-volume ratio, have led to significant changes in their physical, chemical, and electronic behavior compared to bulk materials. Interesting aspects of nanoparticles include size- and shape-dependent interatomic bond distances, melting points, chemical reactivity, and optical and electronic properties. At the nanoscale, a significant fraction of atoms is located on the particle surface, which leads to an increase in the number of active sites and a significant improvement in catalytic activity and charge transfer. In addition, quantum effects in the structure of nanoparticles cause changes in the density of electronic states and improve surface reactivity, which plays an important role in facilitating redox reactions and improving the selectivity of processes.

In the field of electrochemistry, nanoparticles not only improve the performance of electrocatalysts but also enable the development of electrodes and sensors with high sensitivity and stability. Nanoparticles offer performance beyond other materials in systems such as fuel cells, supercapacitors, advanced batteries, and biosensors. The ability to engineer the shape, size, and crystal structure of nanoparticles, along

DOI: 10.1201/9781003229032-2

with the possibility of functionalizing their surfaces, allows researchers to optimize electrochemical properties in a targeted manner. Also, the use of nanoparticles at the scale of nanoelectrodes accelerates mass transfer and enhances the electrode response in fast reactions. Due to these advantages, nanoparticles have become one of the main pillars of recent advances in electrochemistry and are expected to show better performance in electrochemical applications with the development of synthesis methods (1).

2.1.2 Green Synthesis as a Sustainable Alternative

Green synthesis is the design of chemical products and processes that reduce or eliminate the use and production of hazardous materials. Fewer hazardous materials mean less waste and a healthier environment. The cost of waste disposal is also high. Paul Anastas of the United States Environmental Protection Agency has formulated a few simple rules for how to achieve sustainability in chemical manufacturing, called the "Green Chemistry Principles." The effort to improve the environment has led to manufacturing routes that include environmentally friendly protocols. These methods eliminate the high production costs and energy consumption associated with conventional methods of manufacturing nanoparticles using conventional techniques that use hazardous chemicals, such as sodium borohydride or hydrazine. Green manufacturing also involves eliminating complex manufacturing routes, choosing water as the solvent, and replacing toxic reducing agents and coating reagents with renewable and sustainable materials. All these are environmentally friendly and are considered as alternative methods for preparing nanoparticles for various applications. Integrating green chemistry principles into nanotechnology is essential in cases where the development of nanotechnology can benefit from a greener approach that leads to a clean, reliable, environmentally friendly, and low-risk process for preparing nanoparticles. Green preparation of nanoparticles has received much attention in nanotechnology, which is the intersection of nanotechnology and green chemistry. Green preparation of nanoparticles has received much attention compared to physical and chemical preparation because it is more environmentally friendly and has fewer side effects. In particular, the use of plant extracts is the main method of green preparation, which has the advantage that plants are widely distributed and easily available, are also much safer in terms of control, and serve as a source of several metabolites. In green preparation, there is no need for high pressure, energy, temperature, and toxic chemicals (2).

2.2 FUNDAMENTALS OF NANOPARTICLE BIOSYNTHESIS

2.2.1 Biological Agents for NP Synthesis

Biological agents for the synthesis of nanoparticles include the following three categories:

1. *Plant extracts.* Rich in secondary metabolites (flavonoids, alkaloids, terpenoids), which act as reducing and capping agents.
2. *Microorganisms.* Microorganisms can reduce metal ions into nanoparticles (reducing agents).

3. *Enzymes and proteins.* Can mediate the nucleation and growth of nanoparticles and act as stabilizing and coating agents.

The following is a review of each of these biological agents for the synthesis of nanoparticles and the various applications of these biosynthesized nanoparticles in electrochemistry.

2.2.1.1 Biosynthesis of Nanomaterials Using Plants

Plants are nature's chemical factories. Availability and diversity are important features of plants. Plant parts used in the preparation of nanoparticles include leaves, roots, flowers, bark, stems, sap, fruits, and seeds. Each plant has a unique set of phytochemicals that include vitamins, proteins, terpenoids, alkaloids, flavonoids, tannins, polyphenols, sugars, amino acids, organic acids, biopolymers, etc. These secondary metabolites effectively act as reducing and capping agents. Plant techniques can be used for the synthesis of nanoparticles, which can be classified into three methods: intracellular synthesis in plants, extracellular synthesis using plant extracts, and the use of individual phytochemicals. Several plant species have a remarkable ability to accumulate various metals in their cells and subsequently convert these accumulated metals into nanoparticles through intracellular processes. Following the principles of green chemistry, the plant-mediated preparation of nanoparticles has the advantages of simplicity, cheapness, and non-toxicity. Studies on a number of plant-derived reduced nanoparticles have proven that these nanoparticles are quite stable due to their crystalline nature. Green preparation has brought about a huge change in the biomedical field since it uses natural resources for the preparation and stabilization processes. Many successful reports on the preparation of nanoparticles using plant extracts have been found in the literature. Reduced nanoparticles using plant extracts are clean and do not require a multi-step preparation process and can be easily scaled up to larger systems. The plant-mediated preparation of metal and metal oxide nanoparticles, especially silver, gold, palladium, zinc, copper, tin, cadmium, titanium, and zirconium, has attracted much attention due to its simple approach, long-term stability, rapid preparation, and environmental friendliness (3). The following is a review of the various categories of secondary metabolites that effectively act as reducing and capping agents.

2.2.1.1.1 Flavonoids

Flavonoids are a group of polyphenolic compounds that are well-known for their antioxidant properties. Flavonoids are found in many parts of plants, including leaves, seeds, roots, barks, and flowers of various trees, and their extracts are used in the synthesis of bionanoparticles. During the synthesis of nanoparticles, flavonoid-based plant extracts primarily act as reducing agents. They play a pivotal role in this process through several mechanisms. One of these mechanisms is as follows: reduction of metal ions by donating electrons to metal ions and converting them from the ionic form to the metallic state (due to hydroxyl (-OH) groups). This reduction reaction involves the formation of flavonoid-metal ion complexes that enhance electron transfer. For example, quercetin and catechins have been identified as effective reducing agents in the synthesis of silver nanoparticles. Flavonoid-based plant

TABLE 2.1
Synthesis of Nanomaterials Based on Flavonoids

Flavonoid Type	Source	Nanoparticles	Morphology	Ref.
Quercetin (flavonol)	*Hypericum perforatum L.* leaf extract	AgNPs	Spherical, 20–50 nm	(5)
Myrecitin (flavonol)	Plant-derived purified chemical	AuNPs	Spherical, 30–66 nm	(6)
Kaempferol (flavonol)	*Lotus leguminosae* leaf extract	AuNPs	Spherical, ~37 nm	(7)
Rutin (flavonol)	Plant-derived purified chemical	AgNPs	Spherical, 80–85 nm	(8)
Morin (flavonol)	Plant-derived purified chemical	AuNPs	Spherical, 20 ± 6 nm	(9)
Luteolin (flavone)	Plant-derived purified chemical	TiO_2NPs()	Rod-shaped, L 33.3–135 nm × W 16.6–58 nm	(10)
Apigenin (flavone)	Plant-derived purified chemical	AuNPs	Spherical, 14.5 ± 3.6 nm	(11)
Tangeretin (flavone)	Plant-derived purified chemical	PtNPs	Spherical, 10–22 nm	(12)
Nobiletin (flavone)	*Evolvulus alsinoides* leaf extract	PdNPs	Spherical, 20 nm	(13)
Baicalein (flavone)	Plant-derived purified chemical	AuNPs	Spherical, 26.5 nm	(14)
Hesperidin (flavonone)	Plant-derived purified chemical	Hesp-AuNPs	Spherical, 10 to 50 nm	(15)
Naringenin (flavonone)	Plant-derived purified chemical	AgNPs	Spherical, 16.3±4.4 nm	(16)
Genistein (isoflavonoid)	Plant-derived purified chemical	AuNPs	Spherical, 14–33 nm	(17)
Daidzein (isoflavonoid)	Plant-derived purified chemical	AuNPs	Spherical, 9–16 nm	(18)
Catechins (flavanonol)	Plant-derived purified chemical	AuNPs	Polymorphic, with shell size 5.41 ± 0.21 nm	(19)
Catechins (flavanonol)	Plant-derived purified chemical	CuNPs	Spherical, ~10 nm	(20)
Cyanidin (anthocyanin)	*Hibiscus sabdariffa L*	AgNPs	Polymorphic, 26.5 nm	(21)
Delphinidin (anthocyanin)	*Ocimum tenuiflorum* inflorescence extract	AgNPs and SeNPs	Spherical, ~ 65 nm	(22)
Malvidin, petunidin, delphinidin, peonidin (an anthocyanin)	*Bauhinia variegata* flower extract	AgNPs	Spherical, 5–15 nm	(23)
Chalcones	*Helichrysum foetidum*	AuNPs	Spherical, 5–40 nm	(24)
–	Pecan nutshell (*Carya illinoinensis*)	Nb_2O_5 – NPs	Spherical	(25)
–	Citrus extracts	SeNPs	70–250 nm	(26)

TABLE 2.1 (*Continued*)
Synthesis of Nanomaterials Based on Flavonoids

Flavonoid Type	Source	Nanoparticles	Morphology	Ref.
Rutin (flavonol)	*Artemisia annua* hairy root extracts	AgNPs AuNPs	Spherical, 10.9 nm Spherical, 15–40 nm	(27)
Quercetin	*Pogonatherum crinitum* extracts	AgNPs ZnO NPs	Spherical, 9.71–12.32 nm Hexagonal, 18.20–23.75 nm	(28)
–	*Circaea lutetiana Ethanolic extract*	AgNPs	Spherical, 3.8 nm	(29)
–	*Anchusa officinalis*	AgNPs	28.5 nm	(30)
Rutin (flavonol)	*Euterpe edulis* fruits (AnthocyanOx®)	AgNPs	59–78 nm	(31)
–	Capped chestnut shell extract	AgNPs	15 nm	(32)

extracts act secondarily as stabilizing agents. This stabilization prevents aggregation and ensures good dispersion of nanoparticles in solution. In addition, flavonoids can produce biocompatible and low-toxicity nanoparticles. The size and morphology of the synthesized nanoparticles depend on the flavonoid concentration, with higher concentrations facilitating rapid nucleation and minimizing growth, resulting in smaller and more uniform particles (4). The following table reports examples of green synthesis of flavonoid-based nanomaterials. Flavonoids act as reducing and capping agents in these syntheses.

2.2.1.1.2 Alkaloids

The shape and size of bionanoparticles produced by plant extracts depend on the content of the plant extract, reaction time, temperature, pH, etc. The constituent compounds in plant extracts, including alkaloids, polyphenols, terpenoids, phenolic acids, proteins, polysaccharides, vitamins, enzymes, proteins, and organic acids, may reduce metal ions. In this regard, in the past few years, laboratory technologies have been successfully developed that use plant extracts to reduce metal ions and produce nanoparticles. Alkaloids are natural compounds that contain basic nitrogen atoms and are well-known for their pharmacological effects. Alkaloids extracted from many parts of plants, including leaves, seeds, roots, bark, and flowers, are involved in the synthesis of bionanoparticles. In the synthesis of bionanoparticles (gold, silver, platinum, and palladium), alkaloid-based plant extracts may act as capping and reducing agents. Alkaloids such as caffeine and nicotine can reduce metal ions by donating electrons through their nitrogen-containing structures. This electron transfer mechanism facilitates the conversion of metal ions into nanoparticles. The structures of the alkaloids used for biosynthesis affect the shape and size of the resulting nanoparticles. Alkaloid-based plant extracts also have antioxidant, antibacterial, anti-inflammatory, antifungal, antidiabetic, anti-HIV, snake venom–neutralizing,

TABLE 2.2
Synthesis of Nanomaterials Based on Alkaloids

Source	Nanoparticles	Morphology	Ref.
Emblica officinalis fruit extract	AgNPs	Spherical, 15 nm	(33)
Trachyspermum ammi (L.) Sprague	AgNPs	–	(34)
Peganum harmala L. seed extract	AgNPs	Spherical, 22.5–66.2 nm	(35)
Peganum harmala seed alkaloid fraction	Pt and Pd NPs	Spherical (PtNPs ~20 nm), (Pt-PdNPs ~33 nm), (PdNPs ~22 nm)	(36)
Piper retrofractum Vahl extract	Cu NPs	Spherical, 2–10 nm	(37)
Terminalia catappa aqueous extract	Nd_2O_3 (neodymium oxide)	40–60 nm	(38)
Solanum tuberosum L	AgNPs	39.5–80.3 diameter	(39)
Cordia sebestena L. leaves	AgCuO	Sharp, needle-like clusters, 671.6 ± 37.3 nm	(40)
Papaver somniferum L	AgNPs	Triangular, 87 nm	(34)
Bauhinia tomentosa L. leaves	AuNPs, AgNPs	–	(41)
Zingiber officinale Roscoe	AuNPs	5–15nm	(42)
Andrographis paniculata (Burm.f.) Nees leaves	AgNPs	13–27 nm	(43)
Clinacanthus nutans (Burm.f.) Lindau stem	AgNPs	Spherical, 10–180 nm	(44)
Conocarpus lancifolius Engl	AgNPs	75.50 nm	(45)

and larvicidal properties. Therefore, they can enhance the functional applications of nanoparticles, especially in biomedical fields (4). The following table reports examples of green synthesis of alkaloid-based nanomaterials. Alkaloids act as reducing and capping agents in these syntheses.

2.2.1.1.3 Terpenoids

Terpenoids (isoprenoids) are a class of organic compounds produced by plants that act as reducing agents and stabilizers in the synthesis of bionanoparticles. Terpenoids are found in all parts of plants and perform several essential functions. Terpenes are the main and structurally diverse secondary metabolites. Terpenoids primarily act as reducing agents, reducing metal ions by donating electrons (e.g., limonene and geraniol). Their hydrocarbon structures and functional groups facilitate the interaction with metal ions in green synthesis. Alkaloid-based plant extracts secondarily act as stabilizers by acting as capping agents (forming a protective layer around nanoparticles). This stabilization prevents aggregation and increases the dispersion of nanoparticles in solutions and is important for maintaining the size and shape of nanoparticles. In addition, terpenoids are biocompatible and generally recognized as safe (GRAS) compounds; thus, they can enhance the functional applications of

TABLE 2.3
Synthesis of Nanomaterials Based on Terpenes and Terpenoids

Phytochemical Compounds	Source	Nanoparticles	Morphology	Ref.
Flavonoids, tannins, saponins, anthraquinones, terpenoids, glycosides and steroids	*Abutilon indicum L*	ZnNP	50–500 nm, spherical	(47)
Flavonones and terpenoids	*Acacia farnesiana* (sweet acacia) seed extract	AgNP	15–20 nm, spherical	(48)
Flavonoids and terpenoids	*Aegle marmelos (Bael)* fruit extract	AgNP	34.7 nm, spherical	(49)
Flavonoids and terpenoids	*Aerva lanata* leaf extract	AgNP	37–47 nm, spherical	(50)
Terpenoids, flavonoids, and proteins	*Alstonia macrophylla* leaf extract	AgNP	<100 nm, spherical	(51)
Flavonols, triterpenoids, steroids, and tannins	*Alternanthera sessilis*	AgNP	32.81 nm, spherical	(52)
Terpenoids	*Andrographis paniculata* leaves	ZnONP	20.23 nm, hexagonal	(53)
Phenols, terpenoids	*Anthriscus cerefolium (L.) Hoffm.* (Apiaceae)	AgNP	15–20 nm, spherical	(54)
Alkaloids, flavonoids, terpenoids, tannins, saponins, and glucosides	*Artocarpus lakoocha* (fruit and leaves)	AuNP	2.75–3.5 nm, hexagonal, triangular, spherical, rectangular, rhombohedral, and irregular	(55)
Protein, flavanones, and terpenoids	*Azadirachta indica* leaf extract	AgNP	10–43 nm, irregular, spherical	(56)
Terpenoids	*Bacopa monniera* whole plant	AgNP	10 nm, spherical	(57)
Terpenoids and tannins	*Caesalpinia coriaria* leaf extracts	AgNP	40–52 nm, triangular and spherical	(58)
Terpenoids	Chamomile	AgNP	2–12 nm, spherical	(59)
Terpenoids, including linalool, eugenol, and methyl chavicol	*Cinnamon zeylanicum* bark extract	AgNP	31–40 nm, spherical	(60)

nanoparticles, especially in biomedical fields (46). The following table reports examples of green synthesis of terpenoid-based nanomaterials. Terpenoids act as reducing and stabilizing agents in these syntheses.

2.2.1.2 Biosynthesis of Nanomaterials Using Microorganisms

Synthesis using microorganisms is one of the green synthesis methods used to prepare nanoparticles. Nanoparticle synthesis is classified into two routes: extracellular and intracellular. The *extracellular* method involves adding a metal ion solution to the cultured microorganisms, which results in a rapid change in the color of the culture medium, indicating the production of nanoparticles. Here, the most important factors affecting the preparation are the concentration of the metal ion solution, the amount of microorganisms used, and the reaction conditions, such as temperature, pH, and reaction time, which affect the size, shape, and crystalline properties of the nanoparticles. The preparation of nanoparticles using various microorganisms is a time-consuming method and requires a sterile environment due to the high probability of contamination. The main advantage of these nanoparticles is their suitability for biological applications. Noble metal nanoparticles prepared using this method have many biomedical applications (61). The following is a review of the various categories of microorganisms that effectively act as reducing agents.

2.2.1.2.1 Microalgae

The use of algae, especially microalgae, has been considered promising in the biosynthesis of nanoparticles, which is a growing field called "phyconanotechnology." Significant advantages, such as cost-effectiveness, availability of algae, and the ability of proteins and peptides present in algae to stabilize nanoparticles, have driven recent studies on the synthesis of algae-mediated nanoparticles. Algae are being used due to their extraordinary ability to adsorb metals and reduce metal ions, their relatively cheap production costs, and most importantly, their potential for the large-scale production of nanoparticles. The capacity of these microbes to survive in extreme weather conditions more successfully than other microorganisms is another attractive feature. They are known as biological nanofactories because they can utilize living and non-living algal biomass to produce nanoparticles. Unlike many other biosynthesis methods, the synthesis of algal nanoparticles requires relatively less time. Metal reduction is carried out by various biomolecules, including polysaccharides, peptides, and pigments. Microalgal nanoparticles have numerous applications in medical treatment and the catalysis of chemical reactions (62). The following table reports examples of green synthesis of nanomaterials based on microalgae.

2.2.1.2.2 Fungal

Fungal-mediated green synthesis of nanoparticles has recently attracted much attention due to its advantages, including environmental friendliness; versatility in obtaining nanoparticles with diverse sizes, shapes, and compositions; ease of handling and cultivation; high stability; biological activity; and high yield. They are biologically superior in the creation of metal and metal oxide nanoparticles because they have a wide range of alkaloids and intracellular enzymes. Compared to bacteria, fungi can produce more nanoparticles. Due to the abundance of reducible compounds such as alkaloids, enzymes, and proteins in their cell surface membranes, fungi have many advantages over other biological organisms. The most likely mechanism for the creation of metal nanoparticles is enzymatic reduction within the fungal cell wall (62). The following table reports examples of green synthesis of nanomaterials based on fungi.

2.2.1.2.3 Yeast

Yeast-mediated synthesis of nanoparticles has advantages, such as cost-effectiveness, environmental friendliness, and faster production rate of nanoparticles compared to other synthesis routes. In addition, the rapid growth of various yeast strains with ease of collection and maintenance in the laboratory and the relatively higher yield of nanoparticles achievable via yeast-mediated synthesis compared to bacterial-mediated synthesis are the main advantages of yeast-mediated synthesis of biogenic nanoparticles (62). The following table reports examples of green synthesis of nanomaterials based on yeast. The biosynthesis of numerous metal nanoparticles, as shown in the table, uses different yeast species.

2.2.1.2.4 Bacteria

Bacteria-mediated green synthesis of silver nanoparticles has been demonstrated with high efficiency and advantages such as cost-effectiveness, biocompatibility, and control over the size, shape, and composition of nanoparticles during incubation with metal salt solutions. Bacteria can reduce metal ions, making them ideal alternatives for the synthesis of nanoparticles. Several bacterial strains are used for the synthesis of metal and other novel nanomaterials. Metal and metal oxide nanoparticles have been extensively synthesized using prokaryotic microorganisms. Bacterial nanoparticle production has been used because the manipulation of microorganisms is very simple (62). A significant number of bacterial strains have been widely used to create bioreducible metal nanoparticles with different shape and size characteristics, and examples of this green synthesis of nanomaterials based on bacteria are reported in the following table.

2.2.1.2.5 Actinomycetes

Actinobacteria have been commonly employed for the synthesis of extracellular enzymes and secondary metabolites. In recent years, they have also gained popularity as a potential candidate for the production of nanoparticles. This is due to their exceptional ability to produce a diverse range of bioactive compounds as well as their high protein content. The process of nanoparticle synthesis in actinobacteria occurs through two pathways: intracellular and extracellular. However, the extracellular reduction pathway is the most widely used due to the multiple scattering that plays an important role and is consequently the pathway that has attracted the most attention in various commercial applications. The results of various studies indicate that bioengineered nanoparticles are more effective in terms of antimicrobial activity compared to nanoparticles synthesized through conventional methods. The enhanced antimicrobial efficacy of nanoparticles produced through biofabrication is attributed to the immobilization and coating of nanoparticles with a variety of biologically active molecules from actinobacteria (63). The following table reports examples of green synthesis of nanomaterials based on actinobacteria.

2.2.1.3 Biosynthesis of Nanomaterials Using Enzymes and Proteins

2.2.1.3.1 Enzymes

Bacteria and fungi produce a variety of enzymes with different functions. Nanoparticle-producing enzymes can be widely cultivated from microorganisms that are inexpensive and easy to grow. Genetic modification of microbial enzymes can improve the catalytic capabilities for nanoparticle synthesis. This flexibility helps customize the properties of synthesized nanoparticles. Many microbial enzymes are robust and

TABLE 2.4
Synthesis of Nanomaterials Based on Microorganisms

Source (Microorganisms)	Nanoparticles	Morphology	Ref.
Microalgae			
Chlorococcum humicola	AgNP	2–16 nm	(64)
Coelastrella aeroterrestrica	AgNP	4–32 nm, hexagonal quasispherical, triangular, rectangular	(65)
Acutodesmus dimorphus	AgNP	2–20 nm, spherical	(66)
Chlorella sp.	AgNP	10–20 nm, spherical	(67)
Coelastrella sp.	AuNPs	70 nm, spherical	(68)
Chlorella vulgaris	AuNPs	40–60 nm, spheroidal or polyhedral and rod-shaped	(69)
Chlorella vulgaris	AuNPs	2–10 nm, spherical	(70)
Rhizoclonium fontinale	AuNPs	16 nm, spherical	(71)
Tetraselmis suecica	AuNPs	51–120 nm, spherical	(72)
Chlorella vulgaris	Pd NPs	Spherical	(73)
Chlorococcum sp.	FeNPs	20–50 nm, spherical	(74)
Fungal			
Aspergillus niger	AgNP	3–30 nm, spherical	(75)
Alternaria alternaria	AgNP	32.5 nm, spherical	(76)
Talaromyces purpureogenus	AgNP	30–60 nm, spherical	(77)
Fusarium semitectum	AuNPs	25 nm, spherical	(78)
Helminthosporum solani	AuNPs	2–70 nm, polydispersed	(79)
Fusarium solani	AuNPs	40–45 nm, needle- and flower-like	(80)
Penicillium chrysogenum	PtNPs	8.5–15 nm, spherical	(81)
Fusarium oxysporum	PtNPs	70–180 nm, rectangular, triangular, spherical, and aggregates	(82)
Aspergillus terreus	ZnO	10–45 nm, spherical	(83)
Alternaria alternata	Fe_2O_3	75–650 nm	(84)
Yeast			
Schizosaccharomyces pombe	CdS NPs	1–1.5 nm, hexagonal	(85)
Saccharomyces cerevisiae	AuNPs	20–100 nm, spherical	(86)
Yarrowia lipolytica	AuNPs	Nanoplates and spheric	(87)
Pichia jadinii	AuNPs	<100 nm, spherical, triangular, and hexagonal	(88)
Magnusiomyces ingens	AuNPs	Plate-shaped: 16–420 nm Spherical-shaped: 10–80 nm	(89)
Magnusiomyces ingens	Se NPs	70–90 nm, spherical	(89)
Rhodotorula mucilaginosa	Se NPs	83 nm and 478 nm, spherical and rod	(90)
Sachharomyces cerevisiae	ZnO	13–20 nm, spherical	(91)
Pichia kudriavzevii	ZnO	10–61 nm, hexagonal	(92)

TABLE 2.4 (*Continued*)
Synthesis of Nanomaterials Based on Microorganisms

Source (Microorganisms)	Nanoparticles	Morphology	Ref.
Saccharomyces cerevisiae	AgNP	10 nm, spherical	(93)
Yarrowia lipolytica	AgNP	50 nm, spherical	(94)
Bacteria			
Pseudomonas deceptionensis	AgNP	10–30 nm, spherical	(95)
Bacillus amyloliquefaciens	AgNP	3–4 nm, cubic/hexagonal	(96)
Escherichia coli, *Exiguobacterium aurantiacumm*, and *Brevundimonas diminuta*	AgNP	5–50 nm, spherical	(97)
Paracoccus haeundaensis BC74171T	AuNPs	20.93 ± 3.46, spherical	(98)
Escherichia coli DH5α	AuNPs	25 nm, spherical	(99)
Shewanella loihica PV-4	CuNPs	10–16 nm	(100)
Pseudomonas fluorescens	CuNPs	49 nm, spherical/hexagonal	(101)
Shewanella loihica	PdNPs	1–12 nm	(102)
Shewanella oneidensis MR–1	PdNPs	10–100 nm	(103)
Actinomycetes			
Streptomyces sp.	AgNP	21–48 nm, spherical	(104)
Streptomyces sp. VITBT7	AgNP	20–70 nm, spherical	(105)
Streptomyces rochei MHM13	AgNP	22–85 nm, spherical	(106)
Amycolatopsis sp. KMN	AuNP	44.4 nm	(107)
Streptomyces sp. NK52	AuNP	10–100 nm, rods, hexagons, triangles, spheres	(108)
Streptomyces MHM38	CuO	1.72–13.49 nm, spherical	(109)
Streptomyces zaomyceticus Oc-5 and *Streptomyces pseudogriseolus* Acv-11	CuO	78–80 nm, spherical	(110)

stable, allowing them to function under a wide range of conditions. This can simplify synthesis and reduce the need for complex substrates. Microbial enzymes produce biocompatible nanoparticles for medical use. Large-scale synthesis of nanoparticles using microbial enzymes is possible due to their easy scalability for industrial production. Microorganisms and enzymes adapt to different environments. Due to their ease of production and scalability, microbial enzymes have attracted much attention for the synthesis of commercial nanoparticles. Microbial enzymes are popular in nanoparticle synthesis due to their diversity, ease of cultivation, genetic manipulation, and adaptability and provide researchers with a versatile and tunable platform for diverse applications. Several enzymes are involved in the production of nanoparticles and act as catalysts for the synthesis process. Despite their essential role in catalysis, these enzymes do not participate in reactions due to their intrinsic properties. These enzymes can increase the rate of nanoparticle synthesis by creating a catalytic effect and lead to faster formation of nanoparticles. They sometimes act as stabilizers and

TABLE 2.6
Synthesis of Nanomaterials Based on Microbial Enzyme

Enzyme	Microbial Source	Nanoparticles	Size (nm)	Ref.
α-amylase	*Aspergillus oryzae*	AgNPs	10–20	(112)
Aspartate protease	*Aspergillus saitoi*	AuNPs	–	(113)
Hydrolase	*Tethya aurantia*	GaNPs	–	(114)
Hydrogenase	*Sulfate-reducing bacteria* (SRB)	PtNPs	–	(113)
Laccase	*Pleurotus ostreatus*	AuNPs	5–50	(115)
Nitrate reductase	*Bacillus licheniformis*	AgNPs	–	(113)
Sulfite reductase	*Thermomonospora* sp.	AuNPs	–	(116)
Cysteine Protease	*Calotropis procera*	Cu	–	(117)
Curcain protease	*Jatropha curcas*	Zn	–	(118)
Peroxidase	*Armoracia rusticana*	Ag, Au	–	(119)
Urease	*Canavalia ensiformis*	Au, Ag, Pt, ZnO	4–60	(120)
Sulfate reductase	*Fusarium oxysporum*	CdNPs	5–20	(121)
Nitrate reductase	*Fusarium oxysporum*	AgNPs	10–25	(122)
Nitrogenase	*Anabaena flos-aquae*	Au, Pd, Pt, AgNPs	3.5–40	(123)
Lysozyme	Hen egg white	AgNPs	10–62	(124)
Nitrate-dependent reductase	*Fusarium oxysporum*	AgNPs	20–50	(125)
Nitroreductase	*Klebsiella pneumonia*	AgNPs	28.2–122	(126)

reducers. The environmental advantages of enzyme-guided nanoparticle synthesis include reduced toxicity, less-hazardous by-products, lower energy consumption, biocompatibility, and renewable resources. This method increases sustainability and reduces environmental impacts (111). The following table lists examples of reported microbial enzyme–based green synthesis of nanomaterials.

2.2.1.3.2 Proteins

Proteins in plant extracts can also play an important role in the biosynthesis of nanoparticles. Proteins play a role in stabilization through coating mechanisms. They can adsorb onto the surface of nanoparticles and form a protective layer that prevents aggregation. The unique structure of proteins allows them to stabilize nanoparticles by providing steric hindrance and electrostatic repulsion. In addition, amino acid residues in proteins can transfer specific functional groups to the surface of nanoparticles, allowing for further modification and enhancement of their properties for targeted applications. Various bioactive compounds present in plant extracts, including flavonoids, phenolic compounds, terpenoids, alkaloids, and proteins, play vital roles in the biosynthesis of nanoparticles. Their ability to reduce metal ions, stabilize nanoparticles, and enhance their properties contributes to the efficiency and effectiveness of the synthesis process (127). The following table reports examples of green synthesis of protein-based nanomaterials.

TABLE 2.7
Synthesis of Nanomaterials Based on Proteins

Proteins Type	Source	Nanoparticles	Morphology	Ref.
Proteins of algae	Brown algae *Macrocystis pyrifera*	CuO-NPs	2–50 nm, spherical	(127)
Protein of fungi	*Rhizopus oryzae* extract	AuNPs	5–65 nm	(128)
Protein of fungi	*Coriolus versicolor*	AuNPs	5–30 nm	(129)
Protein of marine bacterium (stabilized agent)	*Pseudomonas fluorescens*	AgNPs (protein-stabilized AgNPs)	1–10 nm, spherical	(130)
Protein of fungal source	Phytopathogenic fungus *Macrophomina phaseolina (Tassi)*	AgNPs	5–40 nm, spherical	(131)
Proteins of fungal source	*Aspergillus oryzae* NRRL447 exogenous proteins	AgNPs	Spherical	(132)
Proteins of ant	Extract of weaver ant larvae	AgNPs	7.87 ± 2.54 nm, spherical	(133)
Proteins of plant (leaf) (reducing and capping agent)	Leaf extract of *Piper betle L.*	Biogenic silver–protein (core–shell) NPs (protein-coated silver nanoparticles)	–	(134)

2.2.2 Mechanism of Bionanoparticle Synthesis

In one classification, the mechanism of bionanoparticle synthesis includes the following two categories:

1. Reduction of metal ions by biomolecules to form nanoparticles
2. Stabilization and capping by proteins, polysaccharides, or phenolic compounds

It is not possible to determine the mechanism of nanoparticle formation using different compounds present in plant extracts as reducing and capping agents with complete certainty. However, primary and secondary metabolites present in different parts of the plant undergo oxidation and reduction reactions during the preparation of nanoparticles. Reports on the preparation of plant-mediated metal nanoparticles prove the role of water-soluble phenolics in the reduction and capping of nanoparticles. Bioactive secondary metabolites in plants are classified into phenolic acids (gallic acid, ellagic acid, caffeic acid, and protocatechuic acid), flavonoids (chalcones, anthocyanins, flavonols, flavones, flavanones, and isoflavonoids), stilbenes, lignans, and other polyphenols. Phenolic acids and other polyphenolic compounds form complexes with metal ions through their ortho-position hydroxyl groups, and the easy oxidation of these hydroxyl groups leads to the reduction of the metal ions. Plant proteins (cysteine, arginine, tyrosine) also play an important role in the preparation of nanoparticles. The free amino or carboxylate ends of proteins are attached to metal ions, and the coating process is carried out by forming a layer on the metal nanoparticles. In general, the main functional groups that contribute to the reduction and capping of nanoparticles are hydroxyl (-OH), carboxyl (-COOH), carbonyl $(\text{C}=\text{O})$, carbon–nitrogen double bond $(\text{-C}=\text{N-})$, and carbon–carbon double bond $(\text{-C}=\text{C-})$ (135).

2.2.3 Advantages of Biosynthesis

- *Eco-friendly (green chemistry).* In green synthesis, the principles of green chemistry are followed; that is, the use of hazardous chemicals, the number of volatile organic solvents, and the amount of harsh reaction conditions are minimized. Biological agents such as plant extracts, microbial metabolites, or bio-compounds act as both reducing and stabilizing agents. This:
  1. Reduces the carbon footprint and energy consumption
  2. Prevents the production of toxic effluents
  3. Maintains the ecological balance in the environment
  4. Makes the process completely environmentally friendly
- *Cost-effective.* The starting materials in this method are often renewable and inexpensive resources, such as agricultural waste, food industry by-products, or native plants. The synthesis is usually carried out at ambient temperatures and pressures and does not require complex equipment or high energy. On the other hand, since toxic materials are not involved, the costs of purification and waste management are also eliminated. These factors collectively significantly reduce the final price of the nanoparticle.
- *Biocompatible and stable nanoparticles.* Biological agents stabilize nanoparticles by naturally coating them with biomolecules (bio-capping). These coatings, which can include proteins, polysaccharides, or flavonoids:
  1. Prevent particle aggregation and oxidation (high chemical and physical stability)
  2. Increase biocompatibility and reduce toxicity (critical for medical and biological applications)
  3. Ensure long-term product stability under various conditions
- *Scalable for industrial production.* Green synthesis can be easily scaled up from laboratory scale to industrial production due to its simplicity, feasibility at ambient temperature and pressure, and the use of abundant starting materials. Biological processes can be run in large bioreactors or continuous systems, ensuring mass production of nanoparticles with uniform quality. This advantage makes it attractive for the pharmaceutical, environmental, agricultural, and energy industries.

2.2.4 Factors Affecting Green Synthesis of Nanoparticles

The green synthesis of nanoparticles is influenced by various factors that determine their size, shape, stability, and efficiency. These factors include the type and concentration of the precursor metal salt, the concentration of the reducing agent, the reaction temperature, the reaction time, and the plant/biological source used, the stirring speed, and the reaction conditions. In order to obtain nanoparticles with desirable properties, it is essential to control these factors. In other words, proper optimization increases the efficiency of green synthesis of nanoparticles and improves their performance in various applications in electrochemistry (135).

These factors include the following:

- *pH.* pH alters the charge on nanoparticle surfaces and affects nucleation and growth processes. Higher pH generally leads to smaller particles; lower pH can cause aggregation.
- *Temperature.* Higher temperatures accelerate the reduction process and increase the nucleation rate, resulting in smaller and more uniform nanoparticles. However, excessive heat can cause uncontrolled growth and aggregation, affecting the stability and dispersion of the nanoparticles.
- *Concentration.* The amount of metal precursor and reducing agent. Higher precursor concentrations increase nucleation sites, generating smaller particles, but can accelerate agglomeration; reducing agent concentration affects reduction rate and stability.
- *Time.* The reaction time plays a crucial role in the growth and development of nanoparticles. A longer reaction time can lead to the formation of larger nanoparticles with well-defined shapes. However, if the reaction proceeds for too long, it may cause nanoparticle aggregation and size variation (polydispersity), which can negatively impact their stability and overall functionality.

2.3 TYPES OF BIOSYNTHESIZED NANOPARTICLES

The biosynthesis of nanoparticles using biological agents allows the production of a wide range of nanomaterials with diverse chemical compositions and properties. These nanomaterials can be divided into three main groups based on their chemical composition: metal nanoparticles, metal oxide nanoparticles, and carbon nanoparticles.

2.3.1 Metal Nanoparticles (e.g., Ag, Au, Cu, Pt)

Metal nanoparticles are among the most widely used biosynthesized nanomaterials and are used in various fields due to their unique physical, chemical, and biological properties. In the biosynthesis process, metal ions are reduced by biological compounds (such as polyphenols, flavonoids, proteins, and enzymes) and then stabilized by biological agents in the environment (136).

AgNPs (silver): Very strong antibacterial, antifungal, and antiviral properties; applications in medical devices, food packaging, and water purification

AuNPs (gold): High biocompatibility and functionality as drug carriers, imaging agents, and biosensors

CuNPs (copper): Antimicrobial and catalytic activity, but require effective stabilization to prevent oxidation

PtNPs (platinum): Significant catalytic and anticancer activity, especially in the production of targeted drugs

2.3.1.1 Metal Oxide Nanoparticles (e.g., ZnO, Fe_3O_4, TiO_2, MgO, Co_3O_4)

Metal oxide nanoparticles have wide applications in the fields of energy, environment, and medicine due to their combination of structural and functional properties. In biosynthesis, metal precursors are converted to stable metal oxides in the presence of biological agents (137).

ZnO: Semiconductor with strong photocatalytic and antimicrobial properties; applications in sunscreens, gas sensors, and water purification

Fe_3O: Remarkable magnetic properties; used in MRI imaging, magnetic separation, and targeted drug delivery

TiO_2: High photocatalytic performance; used in photovoltaics, antibacterial coatings, and the removal of organic pollutants

MgO: Alkaline and antimicrobial properties; used in food and agricultural industries

Co_3O: Outstanding electrochemical activity; used in batteries and supercapacitors

2.3.2 Carbon-Based Nanomaterials (e.g., Graphene, Carbon Dots)

Biosynthesized carbon nanomaterials using biological sources have gained a special place due to their biocompatibility, chemical stability, and unique optical and electrical properties.

Graphene. A two-dimensional structure of carbon atoms with very high electrical and thermal conductivity. Its biosynthesis is possible using plant extracts and bacteria, and it is used in sensors, supercapacitors, and drug delivery systems.

Carbon dots (CDs). Nanometer carbon dots with strong photoluminescence and tunable emission wavelength used in bioimaging, optical sensing, and optoelectronics. Biological resources such as fruits, vegetables, and agricultural waste are used as their precursors (138).

2.4 CHARACTERIZATION TECHNIQUES

2.4.1 X-Ray Diffraction (XRD)

XRD examines the crystallinity and phase of bionanoparticles. This method provides information about the crystal structure of the material, which is important for understanding the properties of bionanoparticles, such as stability, optical properties, and interactions with other materials (135).

2.4.2 Fourier Transform Infrared (FTIR) Spectroscopy

FTIR spectroscopy examines the functional groups present on the surface of nanoparticles and the chemical bonds between metal ions and biomolecules (from plant extracts, microorganisms, enzymes, and proteins). FTIR identifies surface functional groups (e.g., -OH, -COOH, $-NH_2$) that play a role in the reduction and stabilization of nanoparticles. This technique provides insights into the biomolecular interactions between biomolecules and metal ions, which helps in understanding their role in synthesis (135).

2.4.3 Scanning Electron Microscopy (SEM)

SEM examines the surface morphology and size of nanoparticles at high resolution. SEM images provide insights into the surface shape of nanoparticles. SEM can also provide a direct estimate of the average particle size. It is possible to observe any aggregation or clustering of nanoparticles, which can affect their properties, with SEM (135).

2.4.4 Transmission Electron Microscopy (TEM)

TEM images have higher resolution than SEM images and are used to observe the internal structure and morphology of nanoparticles at the nanoscale. TEM provides very detailed images and can be used to measure the precise size of nanoparticles. TEM can reveal the internal structure and crystallinity of nanoparticles and help confirm whether they are amorphous or crystalline. TEM can be used to analyze the distribution and aggregation of nanoparticles at a much finer scale than SEM (135).

2.4.5 Dynamic Light Scattering (DLS)

The size distribution of nanoparticles in a liquid suspension and their stability are measured by DLS. DLS provides the average size and polydispersity index (PDI), which indicate the uniformity of the size of the nanoparticles. DLS can also be used to track the stability of the nanoparticle dispersion over time by measuring changes in the size distribution (135).

2.4.6 Zeta Potential Analysis

The surface charge of nanoparticles (which affects their stability in suspension) is measured by zeta potential analysis. *Zeta potential* is the potential difference between the surface of the particles and the bulk solution. This potential difference provides information about the electrostatic repulsion between the particles (135).

2.4.7 UV-Vis Spectrophotometry (UV-Vis)

UV-Vis spectroscopy is used for surface plasmon resonance (SPR). UV-Vis is used to track the progress of nanoparticle synthesis and confirms the reduction of metal ions to nanoparticles. The position and shape of the SPR peak can indicate the size and shape of the metal nanoparticles. The intensity of the SPR peak can also indicate the concentration of the nanoparticles in the solution (135).

2.5 ELECTROCHEMICAL APPLICATIONS OF BIOSYNTHESIZED NANOPARTICLES

Nanoparticles are used in almost all scientific and industrial fields, including in electrochemical sensors (for the detection of biomolecules, heavy metals, pesticides), energy storage devices (including supercapacitors, lithium-ion,[1] and sodium-ion batteries),[2] and fuel cells and HER, OER, ORR electrocatalysts. However, they require advanced manufacturing processes that have adverse effects on human health as well as the environment. In this regard, the green synthesis of nanomaterials using biological agents for synthesis is used as a sustainable alternative.

2.5.1 Electrochemical Sensors

Electrochemical (bio)sensors are considered as sensitive diagnostic tools for disease diagnosis as well as the detection of numerous species of pharmaceutical, clinical, industrial, food, and environmental origin which have attracted much attention due to their high sensitivity and selectivity in complex matrices, minimal sample preparation steps, simple design, rapid response, and portability. In the recent past, numerous electrochemical sensors have been developed based on various nanomaterials, such as carbon materials, polymers, and metal nanoparticles. However, recent advances in nanotechnology have raised environmental risks associated with them. Therefore, ensuring the green production of nanomaterials is very essential in preventing adverse environmental impacts on human health. The biosynthesis of nanomaterials using microorganisms, plant extracts and enzymes, and proteins further promotes the development of green electrochemical sensor platforms. Green electrochemical sensors often use biosynthesized nanoparticles based on plant extracts as natural reducing agents in this green synthesis. The development of green electrochemical sensor platforms is one of the most active research areas that minimize the use of hazardous materials. The advantages of these sensors include non-toxicity, low cost, biodegradability, stability, and sensitive surfaces (139). The following table presents the potential of various green nanoparticles as electrode materials in green electrochemical sensors.

2.5.2 Energy Storage Devices

The conventional chemical synthesis of nanoparticles involves the use of chemicals and has been associated with the advantages of simplicity of operation and time saving. However, the chemical reagents used as reducing agents and stabilizers are often hazardous. Also, there are environmental and health concerns due to their advanced production processes. Conventional chemical synthesis has challenges, such as low yield of nanoparticles and high operating costs, due to the high amount of energy

TABLE 2.8
Examples of Green Synthesized Nanomaterials for Electrochemical (Bio)Sensors to Detect Various Analytes

Natural Agent	Electrode	Analyte	Method	Detection Limit	Linear Range	Ref.
Biomolecules and Pharmaceuticals						
E. tereticornis leaf extract	rGO-AuNPs/SPE	L-tryptophan	DPV	0.39 μM	0.5–500 μM	(140)
Ocimum tenuiflorum leaf extract	ZnO NPs/GCE	Glucose	Amp.	0.043 μM	1–8.6 μM	(141)
Gallic acid	RGO/AuNPs	Dopamine	DPV	0.0026 μM	0.01–100.3 μM	(142)
Syzygium aromaticum extract	MnONPs/BCA/gold electrode	p-nitrophenol Hydroquinone	DPV DPV	15.65 μM 0.15 μM	200–550 μM 3–90 μM	(143)
Justicia glauca leaf extract	AgNPs/Activated SPCE	Dopamine	DPV	0.017 μM	0.05–45.35 μM	(144)
Caesalpinia bonducella seed extract	CuONPs-modified PIGE[3]	Riboflavin	SWV	1.04 μM	3.13–56.3 nM	(145)
Azadirachta indica leaves	AgNPs/GCE	Prostate specific antigen	–	1 pg/ml	1 pg/ml–3 μg/ml	(146)
Carica papaya seed extract	MWCNTs/ZnO NPs/GCE	Silymarin	SWV	0.062 mg/L	0.014–0.152 mg/L	(147)
Peach juice extract	ZnO – NDCS/Gox/GCE	Glucose	DPV	6.3 μM	0.2–12 mM	(148)
Cow urine–reducing agent	rGO	Glucose	LSV	1.9019 μM	0.2–10 mM	(149)
Tea leaves extract reducing and stabilizing agent	NC-fMWCNT-CuN/GCE	Acetaminophen Diclofenac sodium	SWV	0.00014 μM 0.00048 μM	0.01–80 μM 0.05–80 μM	(150)
Trigonella foenum-graecum leaf extract	Urs/GS – TiO_2 – NPs/ITO	Urea	CV	0.022 μM	4–90 μM	(151)
Arum italicum leaf extract reducing and stabilizing agent	Ag/rGO/GCE	vitamin B_2 (Riboflavin)	DPV	0.06 nM	2 μM –2.2 nM	(152)
Phoenix dactylifera seed	AgNPs/CPE	Ornidazole	SWV	7.58×10^{-7} M	1×10^{-5} – 1×10^{-3} M	(153)
Leaves of the gotu kola (*Centella asiatica*)	CaO NPs/CPE	Ascorbic acid Uric acid	CV	–	1–5 mM 1–5 mM	(154)
Butea monosperma leaf	CdO-CdS hybrid nanocomposite	Dopamine	DPV	23 μM	–	(155)
Artocarpus heterophyllus leaf extract	Ag-$ZnFe_2O_4$@graphene nanocomposite	Dopamine	LSV	1.6 μM	102–862 μM	(156)

TABLE 2.8 (*Continued*)
Examples of Green Synthesized Nanomaterials for Electrochemical (Bio)Sensors to Detect Various Analytes

Natural Agent	Electrode	Analyte	Method	Detection Limit	Linear Range	Ref.
Aloe vera latex extract	ZnO/CPE	Paracetamol D-glucose	CV	–	1–5 mM	(157)
Tragacanth gum	ZnO/NiO/GCE	Riboflavin	SWV	8.7 nM	10 nM-1100 μM	(158)
White pitaya (*Hylocereus undatus*) extract	*f*-MWCNT-AuNP-Ext/GCE	Hydroxychloroquine	SWV	0.0093 μM	0.03–17 μM	(159)
Almond gum	PVP/ZnO/SPCE	Dopamine	SWV	1.3375 μM	100–600 μM	(160)
Phoenix dactylifera seed extract	AgNPs	Ornidazole	SWV	3.8×10^{-6} M	1×10^{-3} –5×10^{-5} M	(161)
Heavy Metal						
Amylum extract	C-$S_{0.75}$-HT-C_{800}	Cd^{2+} Hg^{2+} Pb^{2+}	SWASV	3.90 nM 3.86 nM 4.91 nM	–	(162)
Honeycomb polypore biomass waste	NaOH – HPBWPCNSs/GT/NFCE	Hg^{2+}	DPV	2.309 nM	0.05–2 μM	(163)
Bosnian seaweed	nZVI-BPC/GCE	Cd^{2+} Pb^{2+}	SWASV	0.1926 nM 0.2082 nM	2.0–50 μg/L	(164)
Kelp derivatives	Kelp-derived carbon material	Pb^{2+} Cd^{2+}	SWASV	53.4 nM 26.5 nM	–	(165)
Walnut shells	Walnut Shell Biomass Carbon	Pb^{2+} Cd^{2+} Hg^{2+} Cu^{2+}	SWV	0.086 nM 0.175 nM 0.246 nM 0.383 nM	–	(166)
Yeast	Sulfur-doped graphene (SG)/ CNT–COOH/MoS_2/yeast/GCE	Pb^{2+}	SWV	2.16×10^{-15} g/L	10^{-6}-10^{-14} g/L	(167)

Pine cone shells	Fluorescent pine cone carbon dot (PCcd)	Cu^{2+}	–	0.005 µg/mL	–	(168)
Cicada shells	Nitrogen self-doped activated porous biochar (NAC)	Cu^{2+}	DPASV	0.3 ng/L	0.001–1000 µg/L	(169)
Sycamore seeds	Bismuth/Biomass derived activated carbon/CPE	Pb^{2+} Cd^{2+}	SWASV	0.13 µg/L 0.08 µg/L	0.5–50 µg/L	(170)
Almond shells	Nitrogen-doped nanoporous carbon (N-NPC)	Pb^{2+}	DPASV	0.7 mg/L	2.0–120 mg/L	(171)
Oyster mushroom (*Pleurotus* sp.)	Fluorescent blue/green carbon dots	Cd^{2+}	–	58.63 µM	–	(172)
Corn flour	Bismuth oxide doped mesoporous carbon nanocomposite	Pb^{2+} Cd^{2+}	SWASV	1.72 pM 1.58 pM	–	(173)
Other Chemical Compounds (Pesticides, Phenolic and . . .)						
Carboxymethyl cellulose	CuNPs/MWCNTs/GCE	Nitrite	Amp.[4]	1.8 µM	5–1,260 µM	(174)
Tea	rGO-AgNPs/GCE	H_2O_2	CPA[5]	0.73 µM	0.002–160 µM	(175)
Green tea extract	rGO/CPE	Sunset yellow	DPV	27 µM	0.05–10 µM	(176)
Garlic extract	GO-AgNPs/GCE	Nitrite	LSV	2.1 µM	10–180 µM	(177)
Terminalia chebula leaf extract	AuNPs/rGO/GCE	Catechol Resorcinol	DPV DPV	0.12 µM 0.78 µM	3–300 µM 15–150 µM	(178)
Orange (*Citrus sinensis*) peel waste aqueous extract	ZnO NPs / GPE[6]	Formaldehyde	DPV	18 µM	0–100 mM	(179)
C. sempervirens pollen extract	AgNPs/GCE	H_2O_2	DPV	0.23 µM	5 µM–2.5 mM	(180)
Ananas comosus (L.) Merr	A-AgNPs/CPE	H_2O_2	–	–	0.1–1,000 µM	(181)
Acacia nilotica leaf extract	AuNPs / SPCE[7]	Nitrite	LSV	0.03 µM	0.1–1,000 µM	(182)
Stevia rebaudiana reducing agent	Ag_2O / ZnOnanocomposite	H_2O_2	LSV	1.36 nM	20–400 µM	(183)

TABLE 2.9
Examples of Green Synthesized Nanomaterials for Supercapacitors

Natural Agent	Electrode	Electrolyte	Specific Capacitance ($F g^{-1}$)	Energy Density ($Wh kg^{-1}$)	Power Density ($W kg^{-1}$)	Ref.
Bengal gram bean extract	BiVO4@C	PVA / KOH	154.6	42.0	571.6	(184)
Potato water extract	$NH_4CoPO_4 \cdot H_2O$ / GO	KOH	–	85.4	1283	(185)
Hibiscus platanifolius flower extract	Au · NPs	KOH	268.63	–	–	(186)
Nyctanthes arbor-tristis flower	AuNPs	KOH	64	–	–	(187)
Castor oil and virgin green leaf (*Anisomeles malabarica*)	Carbon nanospheres	KOH	–	11.83 mWh cm^{-2}	0.3 mW cm^{-2}	(188)
Aloe vera leaf	Calcium magnesium aluminate (CMA)	KOH	79.1	–	–	(189)
Salvia hispanica seed	$Cu_3Mo_2O_9$NPs	KOH	337.1268	–	–	(190)
Nigella sativa seed	LIG^8 + carbon dots (CDs)240	PVA / H_2SO_4	–	7.9023×10^{-5} mWh cm^{-2}	3.657×10^{-3} mW cm^{-2}	(191)
Sideritis vuralii	Carbon quantum dots	KOH	8.26	4.59 mW g^{-1}	3.30 Wh g^{-1}	(192)
Agricultural or textile wastes	Carbon nanohair/CQDs/ACs composite	H_2SO_4	988.8	–	–	(193)
Phyllanthus emblica seed extract	Ce doped ZnO	KOH	338	–	–	(194)
Ocimum sanctum plant	Ce^{3+}doped V_2O_5NPs (5 mol%)	KOH	180.47	9.30	796	(195)
Saw palmetto extract	PPy/MXene@TiO_2/GC	H_2SO_4	337.9	5.0	1,000	(196)
Ocimum sanctum leaf extract	NiO NPs	KOH	100	–	–	(197)
Opuntia ficus-indica leaf extract	NiO NPs	KOH	280	25	–	(198)

Silk of *Zea mays* leaf	NiO NPs	Na_2SO_4	54 C g^{-1}	–	–	(199)
Red flower extract of *Callistemon viminalis*	NiO NPs	KOH	101	–	–	(200)
Flower extract of *Hydrangea paniculata*	NiO NPs	KOH	644	–	–	(201)
Leaf extract of *Avicennia marina*	NiO NPs	NaOH	–	–	–	(202)
Flower extr`act of *Zephyranthes rosea*	NiO NPs	KOH	548.25	–	–	(203)
Guava leaves	NiO NPs	KOH	85.3	–	–	(204)
C. cladosporioides fungi	NiO NPs	KOH	334	–	–	(205)
Opuntia ficus indica dry leaves	NiO NPs	KOH	–	–	–	(206)
Tamarindus indica collection of dried seeds	$NiMnO_3$NPs	KOH	65.45	–	–	(207)
Green tea leaf extract	ZnO Nps	KOH	–	–	–	(208)
Ficus benghalensis leaf extract	ZnO CuO	$NaHCO_3$	6.318 2.144	1.062 0.1904	–	(209)
Tabernaemontana divericata leaf extract	ZnO	H_2SO_4	535	36.456	65.624	(210)
Sucrose	Zn – C	Na_2SO_4	91	32.61	1K	(211)
Eclipta prostrata leaf	ZnO-rGO	KCl	180	–	–	(212)
Cognata leaves	ZnO @PdO/Pd	KOH	178	3.7	3,718	(213)
Foliar extract of *E. cognata*	ZnO-Co_3O_4	KOH	165	4.1	7.5K	(214)
Piper nigrum extract	ZnO / PTFE	KOH	268.5	–	–	(215)

TABLE 2.9 (*Continued*)
Examples of Green Synthesized Nanomaterials for Supercapacitors

Natural Agent	Electrode	Electrolyte	Specific Capacitance ($F\ g^{-1}$)	Energy Density ($Wh\ kg^{-1}$)	Power Density ($W\ kg^{-1}$)	Ref.
Hibiscus Sabdariffa L extract	$ZnO-CeO_2$	KOH	431	–	–	(216)
Sauropus androgynous leaves	$ZnO\text{-}Co_3O_4$	KOH	180	9.54	375.3	(217)
Moringa oleifera plant	CuO	KOH	338	–	–	(218)
Clerodendrum phlomidis aqueous extract	CuO-GO	KOH	82.1	–	–	(219)
Piperine	CuO-rGO	Na_2SO_4	137	14	12,000	(220)
Mimosa pudica leaf extract	$CuFeS_2$	Na_2SO_4	501	–	–	(221)
Sugarcane molasses	CuONPs / AC	Na_2SO_4	245	45.9	2,091.6	(222)
Maize *Zea mays L.* fresh husk	Cu_2O	Na_2SO_4	87	39.3	737	(223)
Mango bark extract	Cu_2O / RGO	KOH	195	37.7	4,114	(224)
Aspalathus linearis natural extract	RuO_2	KOH	750	–	–	(225)
Akarkara (*Anacyclus pyrethrum*)	RuO_2	H_2SO_4	209	–	–	(226)
Akarkara (*Anacyclus pyrethrum*)	RuO_2	H_2SO_4	209	–	–	(226)
Calotropis gigantea plant leaf extract	TiO_2	Na_2SO_4	804	–	–	(227)
Aloe vera and gram seeds	TiO_2 / rGO-NS	H_2SO_4	599.9	83.31	2,243.36	(228)
Ananas comosus crown extract	SnO_2 / $g\text{-}C_3N_4$	KOH	274	39.3	7,425	(229)

required for heating. Currently, in some parts of the world, there is a problem of sufficient energy supply. To this end, current studies have proposed green synthesized nanoparticles as a sustainable alternative to conventional synthesized nanomaterials for application in alleviating the energy shortage problem. One of the fundamental applications of these biosynthesized nanomaterials using biological agents is in energy storage devices (including supercapacitors, lithium-ion, and sodium-ion batteries) (61).

2.5.2.1 Supercapacitors

Compared with conventional batteries, supercapacitors have higher power density and cycle efficiency in addition to the ability to perform up to millions of charge–discharge cycles. Therefore, the potential applications of supercapacitors in electronic devices are advancing. As a result, there is a great focus on using low-cost technology, such as the green synthesis of nanomaterials in the preparation of nanoparticles as electrode materials to develop supercapacitors with fast charging, high efficiency, high temperature resistance, and higher energy storage capacity (61). The following table presents the potential of various green nanoparticles as electrode materials in green supercapacitors.

2.5.2.2 Lithium-Ion Batteries

Recently, with the increasing use of electric vehicles (EVs) and hybrid electric vehicles (HEVs), the market demand for lithium-ion batteries has increased due to their high specific gravity capacities. However, at present, the energy storage performance of lithium-ion batteries is low, which is due to the low capacity and transfer rate of this battery, which is exhibited by graphite (the main commercial anode material). This challenge necessitates the need to improve lithium-ion batteries through the use of nanomaterials. However, nanomaterials have environmental problems. Therefore, the green synthesis of nanomaterials using biological agents for synthesis is being used as a sustainable alternative (61). The following table presents the potential of various green nanoparticles as electrode materials in green lithium-ion batteries.

2.5.2.3 Sodium-Ion Batteries

Due to the widespread use of lithium-ion batteries for energy storage applications in EVs and HEVs, there is a concern about the depletion of lithium. Sodium-ion batteries are a suitable alternative for large-scale energy storage due to their low cost due to their abundance in the Earth's crust (about 1,350 times more than lithium), constant energy, scalability, and non-toxic and environmentally friendly nature. However, the large mass and radius of sodium ions pose a challenge in designing electrode materials with high charge–discharge cyclic stability. For this purpose, the use of nanomaterials as a remediation strategy for this challenge is necessary. However, nanomaterials have environmental problems. Therefore, green synthesis of nanomaterials using biological agents is being investigated as a sustainable alternative (61). The following table presents the potential of various green nanoparticles as electrode materials in green sodium-ion batteries.

TABLE 2.10
Examples of Green Synthesized Nanomaterials for LIBs

Natural Agent	Electrode	Specific Capacity ($mAh\ g^{-1}$)	Test Current ($mA\ g^{-1}$)	Cycles	Ref.
Cotton stalk	Derived carbon materials	271	271	100	(230)
Natural silk waste fabric	$ZnCo_2O_4$ on biochar substrate	416	416	100	(231)
Norway Spruce Bark	Activated biochars	370	370	5,000	(232)
Wheat protein	Coating of Si@NiO nanoflowers	1,098	100	100	(233)
Seaweed	Modified silicon anodes calcination	714	1,000	1,000	(234)
Wheat flour	Nitrogen-doped self-encapsulated biochar	712	1,000	1,000	(235)
Tea-seed shells	Sn nanoparticles embedded into biochar	872	100	110	(236)
Camellia oleifera	Mn_3O_4/biochar	992	200	100	(237)
Crab shell	Antimony-enriched waste	833	1C	500	(238)
Pine nut shells	Carbon nanotube–NiO composite	674	100	100	(239)
Wheat stalk	Highly graphitic carbon nanosheets	502/161	0.1C/10C	500	(240)
Ginkgo leaf	Nitrogen-doped porous carbon	505/198	0.1C/10C	500	(241)
Tea leaf	Hierarchically porous carbon	499	0.2C	200	(242)
Bean dreg	Modified carbon	396	0.1C	100	(243)
Soyabean	Defect-rich N-doped porous carbon	310/275	2C/4C	500	(244)
Sweet potato	Carbon nanoparticles	235/84	0.5C/10C	200	(245)
Wood	Hard carbon	315	0.1C	150	(246)

2.5.3 Electrocatalysis and Fuel Cells

2.5.3.1 HER, OER, ORR Electrocatalysts

Fuel cells and metal–air batteries can convert the chemical energy of fuel into electricity through the oxygen reduction reaction (ORR), overcoming challenges such as dependence on fossil fuels. The ORR reaction occurs at the cathode of fuel cells and metal–air batteries and typically involves the four-electron pathway ($O_2 + 2e^- + 2H^+ \rightarrow 2H_2O$) in acidic media or ($O_2 + 2H_2O + 4e^- \rightarrow 4OH^-$) in alkaline media, which provides the highest energy efficiency. The alternative two-electron pathway ($O_2 + 2e^- + 2H^+ \rightarrow H_2O_2$) leads to the production of hydrogen peroxide, which is undesirable in terms of efficiency and stability. Currently, noble electrocatalysts such as platinum (Pt) and its alloys are recognized as

TABLE 2.11
Examples of Green Synthesized Nanomaterials for SIBs

Natural Agent	Electrode	Specific Capacity (mAh g^{-1})	Test Current (mA g^{-1})	Cycles	Ref.
Xanthium sibiricum fruit	Honeycomb-structured biochar	202	50	200	(247)
Sugarcane	Hard carbon	323	50	500	(248)
Walnut shell	Hard carbon	283	20	100	(249)
Gelatin	Carbon nanosheets	309	200	200	(250)
Cellulose	Ionic liquid–induced	391	100	100	(251)
Onion waste	Nitrogen-doped carbon	152	50	2,000	(252)
Cellulose, chitosan	Nitrogen-doped biomass/reduced graphene oxide	395	100	500	(253)
Cellulose pulps	Wood-derived carbon with selectively introduced C=O groups	330	1 A/g	400	(254)
Lignin	Hard carbon	183	40	100	(255)
Coconut shells	Hard carbon	270	300	100	(256)
Lychee seeds	Mesoporous disordered carbon pyrolysis	225	20	100	(257)
Pine pollen	Honeycomb-like hard carbon	201	50	200	(258)
Hemp haulm	Three-dimensional free-standing hard carbon	156	200	100	(259)

highly active materials for ORR. However, their high cost, scarcity, and sensitivity to catalytic toxicity (CO , methanol) have prevented their widespread use on an industrial scale. Hence, several strategies such as doping with heteroatoms (N , S , P , B), creating porous structures with high specific surface area, and improving electrical conductivity have been considered to develop alternative high-performance electrocatalysts for ORR. In the meantime, the use of green nanomaterials with the aim of reducing or eliminating the consumption and production of hazardous materials is considered as a key route for the development of sustainable electrochemical technologies (260). The following table presents the potential of various green nanoparticles as electrocatalysts in green ORR electrolysis reactions.

An ideal green energy source, hydrogen (H_2), with an energy density of 142 MJ kg^{-1}, can be produced through electrochemical water splitting, addressing challenges such as the depletion of fossil fuels and the increasing energy demand. Electrochemical water splitting involves a hydrogen evolution reaction (HER) at the cathode and an oxygen evolution reaction (OER) at the anode. HER processes involve two electrons in both acidic ($2H^+ + 2e^- \rightarrow H_2$) and alkaline ($2H_2O + 2e^- \rightarrow H_2 + 2OH^-$) environments, while OER processes involve four electrons in both acidic ($2H_2O \rightarrow O_2 + 4H+ + 4e^-$) and alkaline ($4H \rightarrow O_2 + 2H_2O + 4e^-$) environments. Currently, noble electrocatalysts, such as Ru/Ir-based compounds for OER and Pt

TABLE 2.12
Examples of Green Synthesized Nanomaterials for Electrocatalysis (ORR, OER, HER)

Natural Agent	Biocatalyst	App.	Tafel Slope (mV dec^{-1})	Overpotential (mV vs RHE)	CD[9] (mA cm^{-2})	Ref.
ORR						
Amylum extract	C-$S_{0.75}$-HT-C_{800}	ORR	70.1	277	10	(162)
Natural date pulp	Porous carbon microspheres	ORR	53	–	–	(262)
Pandanus odoratissimus leaf aqueous extracts	Fe_3O_4	ORR	126	390 (for water oxidation reaction)	1	(263)
Lotus root	Fe/N co-doped carbon	ORR	–	660 (half-wave potential vs. RHE)	–	(264)
Sargassum sp. extract	PtNPs	ORR	200		6.3	(265)
Persea americana Mill. extract	rGO	ORR	64	−190 (onset potential vs. Ag|AgCl)	–	(266)
Fenugreek seed extract	Si-GQD NCs[10]	ORR	96	−330 (onset potential vs. Ag|AgCl)	2.61	(267)
Bael gum (BG)	Au nanowires	ORR	80	960 (onset potential vs. RHE)	–	(268)
Chilli plants and their fruits	Iron and nickel-based electrocatalysts derived from biochar (Plant-Fe 800)	ORR	–	970 (onset potential vs. RHE) 870 V (half-wave potential vs. RHE)	10	(269)
Lignin	$Co_{1-x}S$@SNFC[11]	ORR	53	840 V (half-wave potential vs. RHE) 306	–	(270)
Posidonia oceanica biomass	Fe-N-C (MgO @ $MgCl_2$)[12]	ORR	–	−0.70 mA (kinetic current at 0.8 V_{RHE})	–	(271)
OER						
Pandanus odoratissimus leaf aqueous extracts	Fe_3O_4	OER	295	390 (for water oxidation reaction)	1	(263)
Lignin	$Co_{1-x}S$@SNFC[13]	OER	53	840 V (half-wave potential vs. RHE) 306	–	(270)
Gelialgas agar gel	Co_3O_4 – CoO	OER	65	337	10	(272)

Lemon juice	$M_xCu_{1-x}O$ (M = La, Ce x = 0.02 – 0.06)	OER	49 60	340 for $La_xCu_{1-x}O$ 337 for $Ce_xCu_{1-x}O$	10	(273)
Psyllium husk (*Plantago Ovata*)	Co_3O_4	OER	71	328	10	(274)
Alkanna tinctoria root extract	MnCo-OOH@GC	OER	99	250	10	(275)
Flaxseed *Linum usitatissimum*)as a polymerizing agent(	$MnCo_2O_4$	OER	82	296	25	(276)
Red seaweed *Rhodophyta*)as a polymerizing agent(	$(Ni,Co)O - Co_3O_4$	OER	66	356	10	(277)
Sage plant (*Ipomoea asarifolia*)	NiO	OER	60.35	307 for NiO300H	10	(278)
Aloe vera leaf extract	NiO_xNPs	OER	95	413	10	(279)
Aloe vera leaf extract	Copper oxide/cobalt oxide	OER	–	390 453	50 100	(280)
HER						
Chilli plants and their fruits	Iron- and nickel-based electrocatalysts derived from biochar (Chilli-Ni 800)	HER	–	410	10	(269)
Biomass gallic acid	P–MoP@C[14]	HER	–	149 in alkaline 162 in acidic solutions	10	(281)
Cyclodextrins	PdO-NPs	HER	25	–750	–	(282)
Cucumis melo var. agrestis fruit juice	Ag/ZnO nanocomposite	HER	62	–572	10	(283)
Parkia speciosa (stinky bean) peel extract	$CoFe_2O_4$ / TiO_2	HER	86.49	594	2	(284)
Lumbricus rubellus worm	DLBS-1033 protein–based	HER	–	–150 (onset potential vs. RHE)	–	(285)

(*Continued*)

TABLE 2.12 (*Continued*)
Examples of Green Synthesized Nanomaterials for Electrocatalysis (ORR, OER, HER)

Natural Agent	Biocatalyst	App.	Tafel Slope (mV dec^{-1})	Overpotential (mV vs RHE)	CD[9] (mA cm^{-2})	Ref.
Water Splitting (OER-HER)						
Soybeans	FeCo / Co_2P / Fe_2P@NPC[15]	OER	85	281 for OER	10	(286)
		HER	158	280 for HER		
Furfural (biomass-derived)	NiFe@BNGC[16]	OER	146	300 for OER	50	(287)
		HER	124	297 for HER		
Mulberry bark (MT-raw)	MT/CoPB[17]	OER	53	310 for OER	10	(288)
		HER	70	−860 for HER		
Amaranthus viridis leaves	AV-Er_2O_3 – doped · -Bi_2O_3	OER	–	455 for OER	10	(289)
		HER		154 for HER		
Green tea and clove	AgNPs	OER	155	1.65 V (vs. Ag/AgCl) for OER	10	(290)
		HER	110.61	−790 (vs. Ag/AgCl) for HER		
Olea ferruginea Royle leaf extract	ZrO_2	OER	71	400 for OER	10	(291)
		HER	272	42 for HER		
Castor and olive oil (capping agents)	Indium sulfide ($In_{2.77}S_4$)	OER	–	346 for OER	1	(292)
		HER		158 for HER		
Honeycomb polypore biomass waste	NaOH_activated porous nanocarbon	OER	75	305 for OER	10	(163)
		HER	89	447 for HER		
Psidium guajava leaves	rGO@ Co_3O_4	OER	31	539 for OER (potential)	200	(293)
		HER	124	714 for HER (potential)	100	

TABLE 2.13
Examples of Green Synthesized Nanomaterials for Fuel Cells

Natural Agent	Catalyst	Cathode/Anode	FC Types	Power Density	Ref.
Vernomia amygdalina leaf extract	α-MnO_2 / NiO / MPGE[18]	Anode	MFC	412.24 mW m^{-2} Peak current response (2.5 mA)	(294)
Vernomia amygdalina leaf extract	α-MnO_2/based PANI/ MPGE	Anode	MFC	426.26 ± 38.89 mW m^{-2} Current density of 2,485.51 ± 397.31 mA m^{-2}	(295)
Mucilage of *Opuntia ficus-indica*	Mucilage-Ag NPs	Anode glucose oxidation	Biofuel cell	25.7 mW cm^{-2} Current density of 85.7 mA cm^{-2}	(296)
Clove extract	Pt NPs	MOR	DMFC	Current density of 5.43 mA cm^{-2} Anodic peak current at 0.6 V	(297)
Moringa oleifera	Fe_3O_4/PANI/MPGE	Anode	MFC	424.51 ± 6.86 mW m^{-2} Open circuit voltage (645 ± 24.50 mV) Current density of 2,475.01 ± 1.23 mA m^{-2}	(298)
Fig leaves	ZrFeO	Support materials for cathode	PEMFC	93 % catalyst retention for ZrFeO over 250 cycles	(299)
E. coli	*E. coli*-Pt/Pd NPs	Cathode	PEMFC	Mass activity (current density) of 2.8 mA mg^{-1}	(300)
Bacterial cellulose	CF/BC/PANI[19]	Anode	MFC	Current density of 0.009 mA cm^{-2}	(301)
Oil palm lignocellulosic	Green GO/ZnO Green GO/TiO_2	Anode	MFC	ZnO : 912 × 10^{-3} Mw m^{-2} TiO_2 : 608 × 10^{-3} mW m^{-2} Current density (ZnO : 70.17 mA m^{-2} and TiO_2 : 65.78 mA m^{-2})	(302)
Bacterial cellulose	Bacterial cellulose doped with P and C	Cathode	MFC	572.16 mW m^{-2}	(303)
Bacterial cellulose	AOBC/c-MWCNTs[20]	–	EBFC[21]	1.897 W m^{-3} Open circuit voltage 0.14 V	(304)

(Continued)

TABLE 2.13 (*Continued*)
Examples of Green Synthesized Nanomaterials for Fuel Cells

Natural Agent	Catalyst	Cathode/Anode	FC Types	Power Density	Ref.
Chlorella pyrenoidosa	Platinum with Fe/N-C	Anode	MFC	$2,740 \pm 160$ mW m^{-2}	(305)
Vitamin-free casein acid	Carbon cloth with Au/NP	Anode	MFC	18.34 mW m^{-2}	(306)
Spirulina platensis	Graphite with AgNPs@AC composite	Cathode	MFC	482.6 mW m^{-2}	(307)
Mixed sulfate-reducing bacteria	Carbon felt with (Fe + O)	Anode	MFC	309 ± 13 mW m^{-2}	(308)
Saccharomyces cerevisiae	Carbon felt with TiO_2	Anode	MFC	25.9 ± 6.2 W m^{-2}	(309)
Vegetable oil	Carbon fiber brushes with Fe/Ni	Anode	MFC	66.4 Mw m^{-3}	(310)
Bananas	Biochar act900	Cathode	MFC	528.2 mW m^{-2}	(311)
Alfalfa leaves	Heteroatom-doped, lamellar-structured carbon	Cathode	MFC	1,328.9 mW m^{-2}	(312)
Green algae (*Spirogyra*)	Superior carbon belts	Anode	MFC	408 mW m^{-2}	(313)
Pinewood lumber	MnO/graphitic biochar	Cathode	MFC	146.7 mW m^{-2}	(315)
Lotus leaves	Hierarchically structured carbon	Cathode	MFC	511/5 mW m^{-2}	(315)
Sewage sludge	Biochar carbonized at 900°C (SS900)	Cathode	MFC	500 mW m^{-2}	(316)
Coffee waste	Activated carbons	Anode	MFC	3,927 mW m^{-2}	(317)
Ceiba pentandra	Hollow fiber	Anode	MFC	1,738 mW m^{-2}	(318)
Eggplant	Hierarchical porous graphitic biochar	Cathode	MFC	667 mW m^{-2}	(319)
Rice husk	Activated carbon	Cathode	MFC	317.7 mW m^{-2}	(320)
Mixed microalgae	Biochar	Cathode	MFC	2.86 mW m^{-2}	(321)

Watermelon rind	N-doped biochar	Cathode	MFC	262 mW m^{-2}	(322)
Sewage sludge/coconut shell	Carbon monoliths	Anode and cathode	MFC	969 mW m^{-2}	(323)
Chitin	CS-rGO-600[22]	Cathode	Biofuel cell	Open-circuit voltage (OCV) of 790 mV potential of ORR 0.176 V; peak current density of 0.995 mA cm^{-2}	(324)
Azadirachta indica (neem leaves) *Rosa centifolia* (rose petals) *Ocimum sanctum* (basil leaves)	FeNP/PGE	–	Biofuel cell	Current density of 2,790.13 mA cm^{-2} Open-circuit voltage (OCV) of 648 mV	(325)
Pseudomonas aeruginosa	AgNPs	Cathode	MFC	Power density of 4.70W m^{-3}	(326)

for HER, are used as highly active electrocatalysts for water splitting. However, the high cost and scarcity of these compounds hinder their use. Hence, several promising strategies (such as tuning the electronic structure, enhancing the metallic property, increasing the electrocatalytic active surface area (ECSA), enhancing the gas evolution behavior, and improving the conductivity of the electrocatalysts) have been employed to achieve high-performance electrocatalysts for HER and OER, which can enhance the performance of the catalysts. However, preparing electrocatalysts with green chemistry approaches for electrochemical water splitting can reduce or eliminate the use or production of hazardous materials, which is highly desirable for achieving green water electrolysis (261). The following table presents the potential of various green nanoparticles as electrode materials in green water electrolysis (OER-HER).

2.5.3.2 Fuel Cells

Fuel cells are electrochemical cells that convert chemical energy into electricity. However, unlike batteries, a continuous supply of fuel is required to sustain the chemical reaction in a fuel cell. Fuel cells are considered as a clean, reliable, efficient, and environmentally friendly energy system that requires a continuous supply of fuels, such as hydrogen, methanol, and ethanol. Among the types of fuel cells, polymer electrolyte membrane fuel cell (PEMFC), direct methanol fuel cell (DMFC), and microbial fuel cell (MFC) have been widely studied and are considered as efficient green energy technologies. The efficiency of a fuel cell can be improved by using advanced electrodes. Green nanomaterials and green technologies have been investigated in fabricating high-performance electrode materials for fuel cells. Platinum-based electrodes have been replaced with green materials and nanocomposites using green manufacturing approaches to achieve environmentally friendly fuel cells. In this regard, membrane electrode assemblies based on ecological and sustainable electrodes and electrolytes have also been designed. In addition, green nanocomposites have been used to form fuel cell electrolyte membranes. Among fuel cells, microbial fuel cells (MFCs) have attracted research attention for incorporating green and sustainable materials (61). The following table presents the potential of various green nanoparticles as electrode materials for fuel cells. Design and interaction with green nanoparticles have led to synergistic effects on morphology, impedance, resistance, power density, current density, electrochemical properties, proton conductivity, and overall efficiency.

2.6 CONCLUSION AND FUTURE PERSPECTIVE

The biosynthesis of nanomaterials using biological agents, such as microorganisms, plant extracts, enzymes, and proteins, offers a sustainable, low-cost, and environmentally friendly approach that reduces the generation of hazardous waste compared to chemical and physical methods. Biochemical mechanisms and molecular pathways in this process play a decisive role in controlling the composition, size, shape, and uniformity of nanoparticles, and their optimization can improve the stability, catalytic activity, and functional performance of materials. In the meantime, a deeper understanding of the role of plant secondary metabolites (flavonoids, alkaloids, and terpenoids) and effective functional groups in the reduction and capping of metals, along with a detailed study of growth conditions, will pave the way for the production of nanoparticles with controlled properties on an industrial scale. A deeper understanding of the role of microorganisms (microalgae, fungi, yeast, bacteria, and actinomycetes) and enzymes and proteins in synthesis is also a prerequisite. However, the need to develop scalable technologies, simplify processes, and increase the efficiency of biosynthesis remains a major challenge. The application of biosynthetic nanomaterials in electrochemistry covers a wide range, from bio-electrochemical sensors and biosensors to energy storage systems (supercapacitors, lithium-ion and sodium-ion batteries), fuel cells, and electrocatalysis of key reactions such as HER, OER, and ORR. In this area, the use of biological precursors and biomass-derived carbons, along with heteroatom doping and porous structure engineering, has improved the performance of electrodes in terms of conductivity, charge storage capacity, and chemical stability. Also, the development of green nanomaterials and biobased composites for electrodes and electrolytes of fuel cells (especially microbial fuel cells) has provided new perspectives in increasing the power density, efficiency, and durability of these systems.

In the future, the focus of research should be on several key axes:

- *In the synthesis domain.* Discovery and engineering of new biological pathways, identification and optimization of effective biomolecules (metabolites, enzymes, and proteins), and development of continuous and scalable processes for the production of nanomaterials with high quality and uniformity.
- *In the application domain.* Design of green hybrid materials by combining multiple functionalities, improving the integration of biosynthetic nanomaterials with advanced electrode architecture, and development of electrochemical systems with high performance and minimal environmental impacts.
- *At a macro level.* Aligning scientific innovations with industrial and environmental requirements, managing nanomaterial waste, and employing life cycle assessment (LCA) to ensure the genuine sustainability of bionanomaterial-based technologies.

Such an approach could pave the way for the next generation of sustainable, low-cost, and high-performance electrochemical systems and establish the position of nanomaterial biosynthesis as a key pillar of energy and environmental technologies.

NOTES

1. LIBs.
2. SIBs.
3. Paraffin-impregnated graphite electrode.
4. Amperometry.
5. Constant potential amperometry.
6. Graphite paste electrode.
7. Screen-printed carbon electrode.
8. Laser-induced graphene.

9 Current density.
10 Silicon-graphene quantum dot nanocomposites.
11 Three-dimensional N and S co-doped carbon nanosheets loaded with cobalt sulfide nanoparticles.
12 Magnesium-templated Fe-N-doped carbon.
13 Three-dimensional N and S co-doped carbon nanosheets loaded with cobalt sulfide nanoparticles.
14 Mesoporous transition metal phosphide.
15 FeCo-rich phosphides-N, P-doped carbon.
16 Nickel-iron nanoparticles–coupled biomass-derived nitrogen-doped graphitic carbon.
17 Biomass-derived, carbon-based, with cobalt phospho-boride.
18 Modified pencil graphite electrode.
19 Carbon fiber–embedded bacterial cellulose/polyaniline nanocomposite.
20 Amidoxime-modified BC/carboxylated multi-walled CNTs.
21 Enzymatic biofuel cell.
22 Nitrogen-doping of the graphene matrix, and 2D multi-wrinkled graphene

REFERENCES

1. Murray RW. Nanoelectrochemistry: Metal nanoparticles, nanoelectrodes, and nanopores. *Chem Rev* [Internet]. 2008 Jul [cited 2025 Aug 13];108(7):2688–720. Available from: https://pubs.acs.org/doi/abs/10.1021/cr068077e
2. Hussain I, Singh NB, Singh A, Singh H, Singh SC. Green synthesis of nanoparticles and its potential application. *Biotechnol Lett* [Internet]. 2016 Apr 1 [cited 2025 Aug 13];38(4):545–60. Available from: https://link.springer.com/article/10.1007/s10529-015-2026-7
3. Husen A. *Secondary metabolites based green synthesis of nanomaterials and their applications* [Internet]. 2023 [cited 2025 Aug 13]. Available from: https://link.springer.com/content/pdf/10.1007/978-981-99-0927-8.pdf
4. Md Ishak NAI, Kamarudin SK, Timmiati SN. Green synthesis of metal and metal oxide nanoparticles via plant extracts: An overview. *Mater Res Express*. 2019;6.
5. Alahmad A, Feldhoff A, Bigall NC, Rusch P, Scheper T, Walter JG. Hypericum perforatum L.-mediated green synthesis of silver nanoparticles exhibiting antioxidant and anticancer activities. *Nanomater* [Internet]. 2021 Feb 14 [cited 2025 Aug 8];11(2):487. Available from: https://www.mdpi.com/2079-4991/11/2/487/htm
6. Mohan UP, Sriram B, Panneerselvam T, Devaraj S, MubarakAli D, Parasuraman P, et al. Utilization of plant-derived Myricetin molecule coupled with ultrasound for the synthesis of gold nanoparticles against breast cancer. *Naunyn Schmiedebergs Arch Pharmacol* [Internet]. 2020 Oct 1 [cited 2025 Aug 8];393(10):1963–76. Available from: https://link.springer.com/article/10.1007/s00210-020-01874-6
7. Oueslati MH, Tahar LB, Harrath AH. Catalytic, antioxidant and anticancer activities of gold nanoparticles synthesized by kaempferol glucoside from Lotus leguminosae. *Arab J Chem* [Internet]. 2018 Jan 1 [cited 2025 Aug 8];13(1):3112–22. Available from: https://arabjchem.org/catalytic-antioxidant-and-anticancer-activities-of-gold-nanoparticles-synthesized-by-kaempferol-glucoside-from-lotus-leguminosae/
8. Kubavat K, Trivedi P, Ansari H, Kongor A, Panchal M, Jain V, et al. Green synthesis of silver nanoparticles using dietary antioxidant rutin and its biological contour. *Beni-Suef Univ J Basic Appl Sci* [Internet]. 2022 Dec 1 [cited 2025 Aug 8];11(1):1–14. Available from: https://bjbas.springeropen.com/articles/10.1186/s43088-022-00297-x
9. Kondath S, Raghavan BS, Anantanarayanan R, Rajaram R. Synthesis and characterisation of morin reduced gold nanoparticles and its cytotoxicity in MCF-7 cells. *Chem Biol*

Interact [Internet]. 2014 Dec 5 [cited 2025 Aug 8];224:78–88. Available from: https://www.sciencedirect.com/science/article/abs/pii/S0009279714002749?via%3Dihub

10. Chahardoli A, Qalekhani F, Shokoohinia Y, Fattahi A. Luteolin mediated synthesis of rod-shaped rutile titanium dioxide nanoparticles: Assay of their biocompatibility. *J Ind Eng Chem* [Internet]. 2022 Jul 25 [cited 2025 Aug 8];111:211–18. Available from: https://www.sciencedirect.com/science/article/abs/pii/S1226086X2200185X?via%3Dihub
11. Rajendran I, Dhandapani H, Anantanarayanan R, Rajaram R. Apigenin mediated gold nanoparticle synthesis and their anti-cancer effect on human epidermoid carcinoma (A431) cells. *RSC Adv* [Internet]. 2015 Jun 8 [cited 2025 Aug 8];5(63):51055–66. Available from: https://pubs.rsc.org/en/content/articlehtml/2015/ra/c5ra04303d
12. Gurunathan S, Jeyaraj M, Kang MH, Kim JH. Tangeretin-assisted platinum nanoparticles enhance the apoptotic properties of doxorubicin: Combination therapy for osteosarcoma treatment. *Nanomater* [Internet]. 2019 Jul 29 [cited 2025 Aug 8];9(8):1089. Available from: https://www.mdpi.com/2079-4991/9/8/1089/htm
13. Gurunathan S, Kang MH, Jeyaraj M, Kim JH. Palladium nanoparticle-induced oxidative stress, endoplasmic reticulum stress, apoptosis, and immunomodulation enhance the biogenesis and release of exosome in human leukemia monocytic cells (THP-1). *Int J Nanomedicine* [Internet]. 2021 Apr 15 [cited 2025 Aug 8];16:2849–77. Available from: https://www.dovepress.com/palladium-nanoparticle-induced-oxidative-stress-endoplasmic-reticulum-peer-reviewed-fulltext-article-IJN
14. Rajkumari J, Busi S, Vasu AC, Reddy P. Facile green synthesis of baicalein fabricated gold nanoparticles and their antibiofilm activity against Pseudomonas aeruginosa PAO1. *Microb Pathog* [Internet]. 2017 Jun 1 [cited 2025 Aug 8];107:261–9. Available from: https://www.sciencedirect.com/science/article/abs/pii/S0882401017302243?via%3Dihub
15. Sulaiman GM, Waheeb HM, Jabir MS, Khazaal SH, Dewir YH, Naidoo Y. Hesperidin loaded on gold nanoparticles as a drug delivery system for a successful biocompatible, anti-cancer, anti-inflammatory and phagocytosis inducer model. *Sci Rep* [Internet]. 2020 Dec 1 [cited 2025 Aug 8];10(1):1–16. Available from: https://www.nature.com/articles/s41598-020-66419-6
16. Shanmuganathan R, Sathishkumar G, Brindhadevi K, Pugazhendhi A. Fabrication of naringenin functionalized-Ag/RGO nanocomposites for potential bactericidal effects. *J Mater Res Technol* [Internet]. 2020 Jul 1 [cited 2025 Aug 8];9(4):7013–19. Available from: https://www.sciencedirect.com/science/article/pii/S2238785419314085?via%3Dihub
17. Stolarczyk EU, Stolarczyk K, Łaszcz M, Kubiszewski M, Maruszak W, Olejarz W, et al. Synthesis and characterization of genistein conjugated with gold nanoparticles and the study of their cytotoxic properties. *Eur J Pharm Sci* [Internet]. 2017 Jan 1 [cited 2025 Aug 8];96:176–85. Available from: https://www.sciencedirect.com/science/article/abs/pii/S0928098716303840?via%3Dihub
18. Shukla R, Nune SK, Chanda N, Katti K, Mekapothula S, Kulkarni RR, et al. Soybeans as a phytochemical reservoir for the production and stabilization of biocompatible gold nanoparticles. *Small* [Internet]. 2008 Sep 1 [cited 2025 Aug 8];4(9):1425–36. DOI:10.1002/smll.200800525. Available from: https://pubmed.ncbi.nlm.nih.gov/18642250/
19. Choi Y, Choi MJ, Cha SH, Kim YS, Cho S, Park Y. Catechin-capped gold nanoparticles: Green synthesis, characterization, and catalytic activity toward 4-nitrophenol reduction. *Nanoscale Res Lett* [Internet]. 2014 Mar 3 [cited 2025 Aug 8];9(1):1–8. Available from: https://link.springer.com/article/10.1186/1556-276X-9-103
20. Podstawczyk D, Pawłowska A, Bastrzyk A, Czeryba M, Oszmiański J. Reactivity of (+)-catechin with copper(II) ions: The green synthesis of size-controlled sub-10 nm copper nanoparticles. *ACS Sustain Chem Eng* [Internet]. 2019 Oct 21 [cited 2025 Aug 8];7(20):17535–43. Available from: https://pubs.acs.org/doi/abs/10.1021/acssuschemeng.9b05078

21. Kalita NK, Ganguli JN. Hibiscus sabdariffa L. leaf extract mediated green synthesis of silver nanoparticles and its use in catalytic reduction of 4-nitrophenol. *Inorg Nano-Metal Chem* [Internet]. 2017 [cited 2025 Aug 8];47(5):788–93. Available from: https://www.tandfonline.com/doi/abs/10.1080/15533174.2016.1218506
22. Olawale F, Ariatti M, Singh M. Ocimum tenuiflorum L mediated green synthesis of silver and selenium nanoparticles: Antioxidant activity, cytotoxicity and density functional theory studies. *Adv Nat Sci Nanosci Nanotechnol* [Internet]. 2022 Mar 25 [cited 2025 Aug 8];13(1):015015. Available from: https://iopscience.iop.org/article/10.1088/2043-6262/ac5d4a
23. Johnson P, Krishnan V, Loganathan C, Govindhan K, Raji V, Sakayanathan P, et al. Rapid biosynthesis of Bauhinia variegata flower extract-mediated silver nanoparticles: An effective antioxidant scavenger and α-amylase inhibitor. *Artif Cells, Nanomedicine Biotechnol* [Internet]. 2018 Oct 3 [cited 2025 Aug 8];46(7):1488–94. Available from: https://www.tandfonline.com/doi/pdf/10.1080/21691401.2017.1374283
24. Omolaja AA, Pearce B, Omoruyia SI, Badmus JA, Ismail E, Marnewick J, et al. The potential of chalcone-capped gold nanoparticles for the management of diabetes mellitus. *Surfac Interfaces* [Internet]. 2021 Aug 1 [cited 2025 Aug 8];25:101251. Available from: https://www.sciencedirect.com/science/article/abs/pii/S246802302100328X?via%3Dihub
25. Brum LFW, da Silva MDCR, dos Santos C, Pavoski G, Espinosa DCR, da Silva WL. Green synthesis of niobium (V) oxide nanoparticles using pecan nutshell (Carya illinoinensis) and evaluation of its antioxidant activity. *Catal Today* [Internet]. 2024 Oct [cited 2025];445:115106. Available from: https://doi.org/10.1016/j.cattod.2024.115106
26. Borowska M, Pszczoła J, Pawlak K, Ruzik L, Ombugadu JN, Trzaskowski M, et al. Microwave-assisted green synthesis of selenium nanoparticles using citrus extracts: Insights into size-controlled formation and surface characteristics. *Colloids Surfaces A Physicochem Eng Asp*. 2025 Mar;725.
27. Bohdanovych T, Kuzema P, Anishchenko V, Duplij V, Kharchuk M, Lyzhniuk V, et al. Comparison of silver and gold nanoparticles green synthesis by Artemisia annua hairy root extracts. *Biol Open*. 2025;14(3).
28. Shrestha DK, Magar AB, Pathak I, Parajuli N, Sharma S, Raj Sharma K. Comparative analysis of biological activities of green-synthesized silver and zinc oxide nanoparticles using Pogonatherum crinitum extracts. *Arab J Chem*. 2025 Jul:1–14.
29. Iskakova Z, Kozhantayeva A, Temirbekova A, Mukhtubayeva S, Bissenova G, Tekebayeva Z, et al. Green synthesis of silver nanoparticles using circaea lutetiana ethanolic extract: Phytochemical profiling, characterization, and antimicrobial evaluation. *Int J Mol Sci*. 2025;26(12):1–24.
30. Keskin C, Aslan S, Baran MF, Baran A, Eftekhari A, Adıcan MT, et al. Green synthesis and characterization of silver nanoparticles using anchusa officinalis: Antimicrobial and cytotoxic potential. *Int J Nanomedicine*. 2025;20:4481–502.
31. Santos VHM dos, Costa MM de O, Granero FO, Figueiredo CCM, Santos HH, Benevides PJC, et al. Green biosynthesis of silver nanoparticles using anthocyanins-rich extract from Euterpe edulis fruits (AnthocyanOx®): Assessment in vitro of antioxidant and antiglycation activities, and in silico anti-aging activity. *Food Bioprod Process*. 2025 Mar;151:189–201.
32. Sezek T, Beğiç N. Green synthesis, chracterization, and antioxidant properties of silver nanoparticles capped chestnut shell extract. *J Food Meas Charact*. 2025:1–10.
33. Ramesh PS, Kokila T, Geetha D. Plant mediated green synthesis and antibacterial activity of silver nanoparticles using Emblica officinalis fruit extract. *Spectrochim Acta Part A Mol Biomol Spectrosc* [Internet]. 2015 May 5 [cited 2025 Aug 8];142:339–43. Available from: https://www.sciencedirect.com/science/article/abs/pii/S1386142515000839?via%3Dihub

34. Vijayaraghavan K, Nalini SPK, Prakash NU, Madhankumar D. One step green synthesis of silver nano/microparticles using extracts of Trachyspermum ammi and Papaver somniferum. *Colloids Surfaces B Biointerfaces* [Internet]. 2012 Jun 1 [cited 2025 Aug 8];94:114–17. Elsevier. Available from: https://www.sciencedirect.com/science/article/pii/S0927776512000513
35. Almadiy AA, Nenaah GE, Shawer DM. Facile synthesis of silver nanoparticles using harmala alkaloids and their insecticidal and growth inhibitory activities against the khapra beetle. *J Pest Sci (2004)* [Internet]. 2018 Mar 1 [cited 2025 Aug 8];91(2):727–37. Available from: https://link.springer.com/article/10.1007/s10340-017-0924-2
36. Fahmy SA, Fawzy IM, Saleh BM, Issa MY, Bakowsky U, Azzazy HMES. Green synthesis of platinum and palladium nanoparticles using Peganum harmala L. Seed alkaloids: Biological and computational studies. *Nanomaterials* [Internet]. 2021 Apr 1 [cited 2025 Aug 8];11(4):965. Available from: https://www.mdpi.com/2079-4991/11/4/965/htm
37. Amaliyah S, Pangesti DP, Masruri M, Sabarudin A, Sumitro SB. Green synthesis and characterization of copper nanoparticles using Piper retrofractum Vahl extract as bioreductor and capping agent. *Heliyon* [Internet]. 2020 Aug 1 [cited 2025 Aug 8];6(8):e04636. Available from: https://www.cell.com/action/showFullText?pii=S2405844020314808
38. Lembang MS, Yulizar Y, Sudirman S, Apriandanu DOB. A facile method for green synthesis of Nd2O3 nanoparticles using aqueous extract of Terminalia catappa leaf. *AIP Conf Proc* [Internet]. 2018 Oct 22 [cited 2025 Aug 8];2023(1):20093. Available from: https://pubs.aip.org/aip/acp/article/2023/1/020093/1006861/A-facile-method-for-green-synthesis-of-Nd2O3
39. Almadiy AA, Nenaah GE. Ecofriendly synthesis of silver nanoparticles using potato steroidal alkaloids and their activity against phytopathogenic fungi. *Brazilian Arch Biol Technol 2018 SciELO Bras* [Internet]. 2020 [cited 2025 Aug 8];13(7). Available from: https://www.scielo.br/j/babt/a/D6t8Bx7mBWMMNZXgSGSqCqG/?lang=en
40. Ravi L, Selvaraj V, Shankar S, Sundar RD, Segaran G, Kumaresan S, Sadhana V, et al. Methylene blue degradation by AgCuO bimetallic nanomaterial, green synthesized using Cordia sebestena leaves. *Research J Pharm Technol* [Internet]. 2020 [cited 2025 Aug 8];13(7). Available from: https://www.indianjournals.com/ijor.aspx?target=ijor:rjpt&volume=13&issue=7&article=012
41. Mukundan D, Mohankumar R, Vasanthakumari R. Comparative study of synthesized silver and gold nanoparticles using leaves extract of Bauhinia tomentosa Linn and their anticancer efficacy. *Bulletin Mater Sci* [Internet]. 2017 Apr 1 [cited 2025 Aug 8];40(2):335–44. Available from: https://link.springer.com/article/10.1007/s12034-017-1376-2
42. Kumar K, Paul W, Sharma CP. Green synthesis of gold nanoparticles with Zingiber officinale extract: Characterization and blood compatibility. *Process Biochem.* Elsevier [Internet]. 2011 [cited 2025 Aug 8]. Available from: https://www.sciencedirect.com/science/article/pii/S1359511311002510
43. Suriyakalaa U, Antony JJ, Suganya S, Siva D, Sukirtha R, Kamalakkannan S et al. Hepatocurative activity of biosynthesized silver nanoparticles fabricated using Andrographis paniculata. *Colloids Surf B Biointerfaces* [Internet]. 2013 [cited 2025 Aug 8]. Available from: https://www.sciencedirect.com/science/article/pii/S0927776512004444
44. Mat Yusuf SN, Che Mood CN, Ahmad NH, Sandai D, Lee CK, Lim V. Optimization of biogenic synthesis of silver nanoparticles from flavonoid-rich Clinacanthus nutans leaf and stem aqueous extracts. *R Soc Open Sci* [Internet]. 2020 Jul 1 [cited 2025 Aug 8];7(7). Available from: https://royalsocietypublishing.org/doi/abs/10.1098/rsos.200065
45. Raheema RH, Shoker RM. Phytochemicals screening and antibacterial activity of silver nanoparticles, phenols and alkaloids extracts of conocarpus lancifolius. *EurAsian J BioSci* [Internet]. 2020 [cited 2025 Aug 8];14:4829–35. Available from: https://www.researchgate.net/publication/345767443_Phytochemicals_screening_and_antibacterial_activity_of_silver_nanoparticles_phenols_and_alkaloids_extracts_of_conocarpus_lancifolius

46. Venkata AL, Sivaram S, Srilakshman G, Muthuraman MS. Review on terpenoid mediated nanoparticles: Significance, mechanism, and biomedical applications. *Adv Nat Sci Nanosci Nanotechnol*. 2022;13(3).
47. Kulkarni BD, Sultana S, Bora M, Dutta I, Paarakh PM, Basappa VA. In vitro cytotoxicity studies of Zn (Zinc) nanoparticles synthesized from abutilon indicum L. against human cervical cancer (HeLa) cell lines. *Pharmacogn J*. 2016;8(2):127–31.
48. Yallappa S, Manjanna J, Peethambar SK, Rajeshwara AN, Satyanarayan ND. Green synthesis of silver nanoparticles using acacia farnesiana (sweet acacia) seed extract under microwave irradiation and their biological assessment. *J Clust Sci* [Internet]. 2013 Dec 11 [cited 2025 Aug 8];24(4):1081–92. Available from: https://link.springer.com/article/10.1007/s10876-013-0599-7
49. Krupa NDA, Raghavan V. Biosynthesis of silver nanoparticles using aegle marmelos (bael) fruit extract and its application to prevent adhesion of bacteria: A strategy to control microfouling. *Bioinorg Chem Appl* [Internet]. 2014 Jan 1 [cited 2025 Aug 8];2014(1):949538. Available from: https://doi.org/10.1155/2014/949538
50. Farjeen MS, Divya SK, Jeyadoss T, Kumar MA. Quorum quenching and antibacterial activity of silver nanoparticles synthesized from medicinal plants against methicillin-resistant Staphylococcus aureus (MRSA). *Int J Pharm Sci* [Internet]. 2014 [cited 2025 Aug 8]. Available from: https://www.researchgate.net/publication/286315718_Quorum_quenching_and_antibacterial_activity_of_silver_nanoparticles_synthesized_from_medicinal_plants_against_Methicillin-resistant_Staphylococcus_aureus_MRSA
51. Borase HP, Patil CD, Salunkhe RB, Suryawanshi RK, Salunke BK, Patil SV. Catalytic and synergistic antibacterial potential of green synthesized silver nanoparticles: Their ecotoxicological evaluation on Poecillia reticulata. *Biotechnol Appl Biochem* [Internet]. 2014 Jul 1 [cited 2025 Aug 8];61(4):385–94. Available from: https://doi.org/10.1002/bab.1189
52. Firdhouse MJ, Lalitha P. Biosynthesis of silver nanoparticles using the extract of Alternanthera sessilis-antiproliferative effect against prostate cancer cells. *Cancer Nanotechnol* [Internet]. 2013 Dec 15 [cited 2025 Aug 8];4(6):137–43. Available from: https://cancer-nano.biomedcentral.com/articles/10.1007/s12645-013-0045-4
53. Kavitha S, Dhamodaran M, Prasad R, Ganesan M. Synthesis and characterisation of zinc oxide nanoparticles using terpenoid fractions of Andrographis paniculata leaves. *Int Nano Lett* [Internet]. 2017 Apr 19 [cited 2025 Aug 8];7(2 (June 2017)):141–7. Available from: https://oiccpress.com/inl/article/view/9014/10050
54. Ortan A, Fierascu I, Ungureanu C, Fierascu RC, Avramescu SM, Dumitrescu O, et al. Innovative phytosynthesized silver nanoarchitectures with enhanced antifungal and antioxidant properties. *Appl Surf Sci* [Internet]. 2015 Dec 15 [cited 2025 Aug 8];358:540–8. Available from: https://www.sciencedirect.com/science/article/abs/pii/S0169433215017274?via%3Dihub
55. Sharma A, Dhiman N, Singh BP, Gathania AK. Green synthesis of gold nanoparticles using extracts of Artocarpus Lakoocha fruit and its leaves, and Eriobotrya Japonica leaves. *Mater Res Express* [Internet]. 2014 Jun 19 [cited 2025 Aug 8];1(2):025042. Available from: https://iopscience.iop.org/article/10.1088/2053-1591/1/2/025042
56. Balachandran YL, Peranantham P, Selvakumar R, Gutleb AC, Girija S. Size-controlled green synthesis of silver nanoparticles using dual functional plant leaf extract at room temperature. *Int J Green Nanotechnol Biomed* [Internet]. 2012 [cited 2025 Aug 8];4(3):310–25. Available from: https://www.tandfonline.com/doi/abs/10.1080/19430892.2012.706183
57. Mahitha B, Raju BDP, Dillip GR, Reddy C, Mallikarjuna K, Manoj L, et al. Biosynthesis, characterization and antimicrobial studies of AgNPs extract from Bacopa monniera whole plant. *Digest J Nanomater Biost* [Internet]. 2011 [cited 2025 Aug 8]. Available from: https://chalcogen.ro/135_Mahitha.pdf

58. Jeeva K, Thiyagarajan M, Elangovan V, Geetha N, Venkatachalam P. Caesalpinia coriaria leaf extracts mediated biosynthesis of metallic silver nanoparticles and their antibacterial activity against clinically isolated pathogens. *Ind Crops Prod* [Internet]. 2014 Jan 1 [cited 2025 Aug 8];52:714–20. Available from: https://www.sciencedirect.com/science/article/abs/pii/S0926669013006638?via%3Dihub
59. Parlinska-Wojtan M, Kus-Liskiewicz M, Depciuch J, Sadik O. Green synthesis and antibacterial effects of aqueous colloidal solutions of silver nanoparticles using camomile terpenoids as a combined reducing and capping agent. *Bioprocess Biosyst Eng* [Internet]. 2016 Aug 1 [cited 2025 Aug 8];39(8):1213–23. Available from: https://link.springer.com/article/10.1007/s00449-016-1599-4
60. Sathishkumar M, Sneha K, Won SW, Cho CW, Kim S, Yun YS. Cinnamon zeylanicum bark extract and powder mediated green synthesis of nano-crystalline silver particles and its bactericidal activity. *Colloids Surf B Biointerfaces* [Internet]. 2009 Oct 15 [cited 2025 Aug 8];73(2):332–8. Available from: https://www.sciencedirect.com/science/article/abs/pii/S0927776509002410?via%3Dihub
61. Akinnawo SO. A review on the green synthesis of nanoparticles for energy conversion and storage. *Kuwait J Sci*. 2025;52(3).
62. Cardoso B, Nobrega G, Afonso IS, Ribeiro JE, Lima RA. Sustainable green synthesis of metallic nanoparticle using plants and microorganisms: A review of biosynthesis methods, mechanisms, toxicity, and applications. *J Environ Chem Eng*. 2025;13(3).
63. Eweis AA, El-Raheem HA, Ahmad MS, Hozzein WN, Mahmoud R. Green fabrication of nanomaterials using microorganisms as nano-factories. *J Clust Sci*. 2024;35(7): 2149–76.
64. Jena J, Pradhan N, Dash BP, Sukla LB, Panda PK. Biosynthesis and characterization of silver nanoparticles using microalga Chlorococcum humicola and its antibacterial activity. *Int J Nanomater Biostruct* [Internet]. 2013 [cited 2025 Aug 8]. Available from: https://www.academia.edu/download/33688047/CHLOROCOCCUM_HUMICOLA_FTIR.pdf
65. Hamida RS, Ali MA, Almohawes ZN, Alahdal H, Momenah MA, Bin-Meferij MM. Green synthesis of hexagonal silver nanoparticles using a novel microalgae coelastrella aeroterrestrica strain BA_Chlo4 and resulting anticancer, antibacterial, and antioxidant activities. *Pharmaceutics* [Internet]. 2022 [cited 2025 Aug 8]. Available from: https://www.mdpi.com/1999-4923/14/10/2002
66. Chokshi K, Pancha I, Ghosh T, Paliwal C, Maurya R, Ghosh A, et al. Green synthesis, characterization and antioxidant potential of silver nanoparticles biosynthesized from de-oiled biomass of thermotolerant oleaginous microalgae Acutodesmus dimorphus. *RSC Adv* [Internet]. 2016 [cited 2025 Aug 8]. Available from: https://pubs.rsc.org/en/content/articlehtml/2016/ra/c6ra15322d
67. Kashyap M, Samadhiya K, Ghosh A, Anand V, Shirage PM, Bala K. Screening of microalgae for biosynthesis and optimization of Ag/AgCl nano hybrids having antibacterial effect. *RSC Adv* [Internet]. 2019 [cited 2025 Aug 8]. Available from: https://pubs.rsc.org/en/content/articlehtml/2019/ra/c9ra04451e
68. MubarakAli D, Arunkumar J, Nag KH, SheikSyedIshack KA, Baldev E, Pandiaraj D, et al. Gold nanoparticles from Pro and eukaryotic photosynthetic microorganisms – comparative studies on synthesis and its application on biolabelling. *Colloids Surf B Biointerfaces* [Internet]. 2013 [cited 2025 Aug 8]. Available from: https://www.sciencedirect.com/science/article/pii/S0927776512005711
69. Luangpipat T, Beattie IR, Chisti Y, Haverkamp RG. Gold nanoparticles produced in a microalga. *J Nanopart Res* [Internet]. 2011 Dec [cited 2025 Aug 8];13(12):6439–45. Available from: https://link.springer.com/article/10.1007/s11051-011-0397-9
70. Annamalai J, Nallamuthu T. Characterization of biosynthesized gold nanoparticles from aqueous extract of Chlorella vulgaris and their anti-pathogenic properties. *Appl*

Nanosci [Internet]. 2015 Jun 1 [cited 2025 Aug 8];5(5):603–7. Available from: https://link.springer.com/article/10.1007/s13204-014-0353-y

71. Parial D, Pal R. Biosynthesis of monodisperse gold nanoparticles by green alga Rhizoclonium and associated biochemical changes. *J Appl Phycol* [Internet]. 2015 Apr 1 [cited 2025 Aug 8];27(2):975–84. Available from: https://link.springer.com/article/10.1007/s10811-014-0355-x
72. Shakibaie M, Forootanfar H, Mollazadeh-Moghaddam K, Bagherzadeh Z, Nafissi-Varcheh N, Shahverdi AR, et al. Green synthesis of gold nanoparticles by the marine microalga Tetraselmis suecica. *Biotechnol Appl Biochem* [Internet]. 2010 Oct [cited 2025 Aug 8];57(2):71–5. Available from: https://iubmb.onlinelibrary.wiley.com/doi/abs/10.1042/BA20100196?casa_token=C_MRvjXZ3tkAAAAA:Cbwfe4eToYAM-76WO2Ui2JlvHCiOkFWVMoybZOEA-s1ReaDDOsmKevJFTUOdhadT0oatGt-kqj_kGqeE
73. Eroglu E, Chen X, Bradshaw M, Agarwal V, Zou J, Stewart SG, et al. Biogenic production of palladium nanocrystals using microalgae and their immobilization on chitosan nanofibers for catalytic applications. *RSC Adv* [Internet]. 2013 [cited 2025 Aug 8]. Available from: https://pubs.rsc.org/en/content/articlehtml/2013/ra/c2ra22402j
74. Subramaniyam V, Subashchandrabose SR, Thavamani P, Megharaj M, Chen Z, Naidu R. Chlorococcum sp. MM11 – a novel phyco-nanofactory for the synthesis of iron nanoparticles. *J Appl Phycol* [Internet]. 2015 Oct 30 [cited 2025 Aug 8];27(5):1861–9. Available from: https://link.springer.com/article/10.1007/s10811-014-0492-2
75. Jaidev LR, Narasimha G. Fungal mediated biosynthesis of silver nanoparticles, characterization and antimicrobial activity. *Colloids Surf B Biointerfaces* [Internet]. 2010 [cited 2025 Aug 8];81:430–3. Available from: https://www.sciencedirect.com/science/article/pii/S0927776510004133
76. Gajbhiye M, Kesharwani J, Ingle A, Gade A, Rai M. Fungus-mediated synthesis of silver nanoparticles and their activity against pathogenic fungi in combination with fluconazole. *Nanomed Nanotechnol Biol Med* [Internet]. 2009 [cited 2025 Aug 8]. Available from: https://www.sciencedirect.com/science/article/pii/S1549963409001129
77. Sharma A, Sagar A, Rana J, Rani R. Green synthesis of silver nanoparticles and its antibacterial activity using fungus Talaromyces purpureogenus isolated from Taxus baccata Linn. *Micro Nano Syst Lett* [Internet]. 2022 Dec 1 [cited 2025 Aug 8];10(1). Available from: https://link.springer.com/article/10.1186/s40486-022-00144-9
78. Sawle BD, Salimath B, Deshpande R, Bedre MD, Prabhakar BK, Venkataraman A. Biosynthesis and stabilization of Au and Au-Ag alloy nanoparticles by fungus, Fusarium semitectum. *Sci Technol Adv Mater* [Internet]. 2008 Jul 1 [cited 2025 Aug 8];9(3). Available from: https://www.tandfonline.com/doi/full/10.1088/1468-6996/9/3/035012
79. Kumar SA, Peter YA, Nadeau JL. Facile biosynthesis, separation and conjugation of gold nanoparticles to doxorubicin. *Nanotechnology* [Internet]. 2008 [cited 2025 Aug 8]. Available from: https://iopscience.iop.org/article/10.1088/0957-4484/19/49/495101/meta
80. Clarance P, Luvankar B, Sales J, Khusro A, Agastian P, Tack JC, Al Khulaifi MM, Al-Shwaiman HA, Elgorban AM, Syed A, Kim HJ. Green synthesis and characterization of gold nanoparticles using endophytic fungi Fusarium solani and its in-vitro anticancer and and biomedical applications. *Saudi J Biol Sci* [Internet]. [cited 2025 Aug 8]. Available from: https://www.sciencedirect.com/science/article/pii/S1319562X19303262
81. Subramaniyan SA, Sheet S, Vinothkannan M, Yoo DJ, Lee YS, Belal SA, Shim KS. One-pot facile synthesis of Pt nanoparticles using cultural filtrate of microgravity simulated grown P. chrysogenum and their activity on bacteria and cancer cells. *J Nanosci Nanotechnol* [Internet]. 2018 [cited 2025 Aug 8]. Available from: https://www.ingentaconnect.com/contentone/asp/jnn/2018/00000018/00000005/art00009
82. Govender Y, Riddin T, Gericke M, Whiteley CG. Bioreduction of platinum salts into nanoparticles: a mechanistic perspective. *Biotechnol Lett* [Internet]. 2009 Jan [cited

2025 Aug 8];31(1):95–100. Available from: https://link.springer.com/article/10.1007/s10529-008-9825-z

83. Fouda A, Saad EL, Salem SS, Shaheen TI. In-Vitro cytotoxicity, antibacterial, and UV protection properties of the biosynthesized Zinc oxide nanoparticles for medical textile applications. *Microb Pathog* [Internet]. 2018 [cited 2025 Aug 8]. Available from: https://www.sciencedirect.com/science/article/pii/S0882401017314985
84. Sarkar J, Mollick MM, Chattopadhyay D, Acharya K. An eco-friendly route of γ-Fe2O3 nanoparticles formation and investigation of the mechanical properties of the HPMC-γ-Fe2O3 nanocomposites. *Bioprocess Biosyst Eng* [Internet]. 2017 Mar 1 [cited 2025 Aug 8];40(3):351–9. Available from: https://link.springer.com/article/10.1007/S00449-016-1702-X
85. Kowshik M, Deshmukh N, Vogel W, Urban J, Kulkarni SK, Paknikar KM. Microbial synthesis of semiconductor CdS nanoparticles, their characterization, and their use in the fabrication of an ideal diode. *Biotechnol Bioeng* [Internet]. 2002 Jun 5 [cited 2025 Aug 8];78(5):583–8. DOI: 10.1002/bit.10233. Available from: https://pubmed.ncbi.nlm.nih.gov/12115128/
86. Lim HA, Mishra A, Yun S Il. Effect of pH on the extra cellular synthesis of gold and silver nanoparticles by Saccharomyces cerevisae. *J Nanosci Nanotechnol*. 2011 Jan;11(1):518–22.
87. Pimprikar PS, Joshi SS, Kumar AR, Zinjarde SS, Kulkarni SK. Influence of biomass and gold salt concentration on nanoparticle synthesis by the tropical marine yeast Yarrowia lipolytica NCIM 3589. *Colloids Surf B Biointerfaces* [Internet]. 2009 Nov 1 [cited 2025 Aug 8];74(1):309–16. Available from: https://www.sciencedirect.com/science/article/pii/S0927776509003373
88. Gericke M, Pinches A. Microbial production of gold nanoparticles. *Gold Bull*. 2006;39(1):22–8.
89. Zhang X, Qu Y, Shen W, Wang J, Li H, Zhang Z, et al. Biogenic synthesis of gold nanoparticles by yeast Magnusiomyces ingens LH-F1 for catalytic reduction of nitrophenols. *Colloids Surf A Physicochem Eng Asp* [Internet]. 2016 May 20 [cited 2025 Aug 8];497:280–5. Available from: https://www.sciencedirect.com/science/article/abs/pii/S0927775716301108?via%3Dihub
90. Ashengroph M, Tozandehjani S. Optimized resting cell method for green synthesis of selenium nanoparticles from a new Rhodotorula mucilaginosa strain. *Process Biochem* [Internet]. 2022 May 1 [cited 2025 Aug 8];116:197–205. Available from: https://www.sciencedirect.com/science/article/abs/pii/S1359511322000964?via%3Dihub
91. El-Khawaga AM, Elsayed MA, Gobara M, Suliman AA, Hashem AH, Zaher AA, et al. Green synthesized ZnO nanoparticles by Saccharomyces cerevisiae and their antibacterial activity and photocatalytic degradation. *Biomass Convers Biorefinery* [Internet]. 2025 Jan 1 [cited 2025 Aug 8];15(2):2673–84. Available from: https://link.springer.com/article/10.1007/s13399-023-04827-0
92. Moghaddam AB, Moniri M, Azizi S, Rahim RA, Ariff A Bin, Saad WZ, et al. Biosynthesis of ZnO nanoparticles by a new pichia kudriavzevii yeast strain and evaluation of their antimicrobial and antioxidant activities. *Mol* [Internet]. 2017 May 24 [cited 2025 Aug 8];22(6):872. Available from: https://www.mdpi.com/1420-3049/22/6/872/htm
93. Roy K, Sarkar CK, Ghosh CK. Photocatalytic activity of biogenic silver nanoparticles synthesized using yeast (Saccharomyces cerevisiae) extract. *Appl Nanosci* [Internet]. 2015 Nov 1 [cited 2025 Aug 8];5(8):953–9. Available from: https://link.springer.com/article/10.1007/s13204-014-0392-4
94. Katharine SD, A T, E PK, Aadhil JMM, Saxena M, Radha P. Sustainable biosynthesis of silver nanoparticles and their application to recover "single cell oil" from yarrowia lipolytica for biodiesel synthesis. *Bionanoscience* [Internet]. 2022 Sep 1 [cited 2025 Aug 8];12(3):890–900. Available from: https://link.springer.com/article/10.1007/s12668-022-00985-z

95. Jo JH, Singh P, Kim YJ, Wang C, Mathiyalagan R, Jin CG, et al. Pseudomonas deceptionensis DC5-mediated synthesis of extracellular silver nanoparticles. *Artif Cells Nanomed Biotechnol* [Internet]. 2016 Aug 17 [cited 2025 Aug 8];44(6):1576–81. Available from: https://www.tandfonline.com/doi/abs/10.3109/21691401.2015.1068792
96. Singh BR, Dwivedi S, Al-Khedhairy AA, Musarrat J. Synthesis of stable cadmium sulfide nanoparticles using surfactin produced by Bacillus amyloliquifaciens strain KSU-109. *Colloids Surf B Biointerfaces* [Internet]. 2011 [cited 2025 Aug 8]. Available from: https://www.sciencedirect.com/science/article/pii/S0927776511001081
97. Saeed S, Iqbal A, Ashraf MA. Bacterial-mediated synthesis of silver nanoparticles and their significant effect against pathogens. *Environ Sci Pollut Res* [Internet]. 2020 Oct 1 [cited 2025 Aug 8];27(30):37347–56. Available from: https://link.springer.com/article/10.1007/s11356-020-07610-0
98. Patil MP, Kang MJ, Niyonizigiye I, Singh A, Kim JO, Seo YB, Kim GD. Extracellular synthesis of gold nanoparticles using the marine bacterium Paracoccus haeundaensis BC74171T and evaluation of their antioxidant activity and antiproliferative effect on normal and cancer cell lines. *Colloids Surf B Biointerfaces* [Internet]. 2019 [cited 2025 Aug 8]. Available from: https://www.sciencedirect.com/science/article/pii/S0927776519305995
99. Du L, Jiang H, Liu X, Wang E. Biosynthesis of gold nanoparticles assisted by Escherichia coli DH5α and its application on direct electrochemistry of hemoglobin. *Electrochem. commun* [Internet]. 2007 [cited 2025 Aug 8]. Available from: https://www.sciencedirect.com/science/article/pii/S1388248107000100
100. Lv Q, Zhang B, Xing X, Zhao Y, Cai R, Wang W, Gu Q. Biosynthesis of copper nanoparticles using Shewanella loihica PV-4 with antibacterial activity: novel approach and mechanisms investigation. *J Hazard Mater* [Internet]. 2018 [cited 2025 Aug 8]. Available from: https://www.sciencedirect.com/science/article/pii/S0304389417309779
101. Shantkriti S, Rani P. Biological synthesis of copper nanoparticles using Pseudomonas fluorescens. *Int J Curr Microbiol App Sci* [Internet]. 2014 [cited 2025 Aug 8]. Available from: https://www.researchgate.net/profile/Shantkriti-Srinivasan/publication/265783874_Biological_synthesis_of_Copper_nanoparticles_using_Pseudomonas_fluorescens/links/57ac243d08ae7a6420c05b00/Biological-synthesis-of-Copper-nanoparticles-using-Pseudomonas-fluorescens.pdf
102. Ahmed E, Kalathil S, Shi L, Alharbi O, Wang P. Synthesis of ultra-small platinum, palladium and gold nanoparticles by Shewanella loihica PV-4 electrochemically active biofilms and their enhanced catalytic activities. *J Saudi Chem Soc* [Internet]. 2018 [cited 2025 Aug 8]. Available from: https://www.sciencedirect.com/science/article/pii/S131961031830019X
103. Wang W, Zhang B, He Z. Bioelectrochemical deposition of palladium nanoparticles as catalysts by Shewanella oneidensis MR-1 towards enhanced hydrogen production in microbial electrolysis cells. *Electrochimica Acta* [Internet]. 2019 [cited 2025 Aug 8]. Available from: https://www.sciencedirect.com/science/article/pii/S0013468619311818
104. Sivalingam P, Antony JJ, Siva D, Achiraman S, Anbarasu K. Mangrove Streptomyces sp. BDUKAS10 as nanofactory for fabrication of bactericidal silver nanoparticles. *Colloids Surf B Biointerfaces* [Internet]. 2012 [cited 2025 Aug 8]. Available from: https://www.sciencedirect.com/science/article/pii/S0927776512002457
105. Subashini JJ, Kannabiran KK. Antimicrobial activity of Streptomyces sp. VITBT7 and its synthesized silver nanoparticles against medically important fungal and bacterial pathogens. *Der Pharmacia Lettre* [Internet]. 2013 [cited 2025 Aug 8]. Available from: https://www.cabidigitallibrary.org/doi/full/10.5555/20133263356
106. Abd-Elnaby HM, Abo-Elala GM, Abdel-Raouf UM, Hamed MM. Antibacterial and anticancer activity of extracellular synthesized silver nanoparticles from marine

Streptomyces rochei MHM13. *Egypt J Aquat Res* [Internet]. 2016 [cited 2025 Aug 8]. Available from: https://www.sciencedirect.com/science/article/pii/S1687428516300279
107. Kabiri F, Aghaei SS, Pourbabaee AA, Soleimani M, Komeili Movahhed T. Antibiofilm and cytotoxic potential of extracellular biosynthesized gold nanoparticles using actinobacteria Amycolatopsis sp. KMN. Taylor Fr Kabiri, SS Aghaei, AA Pourbabaee, M Soleimani, T Komeili MovahhedPreparative Biochem Biotechnol 2023•Taylor Fr [Internet]. 2023 [cited 2025 Aug 8];53(3):265–78. Available from: https://www.tandfonline.com/doi/abs/10.1080/10826068.2022.2076244
108. Prakash D, Mahale V, Bankar A, Nawani N, Zinjarde S, Kapadnis, B. Biosynthesis of colloidal gold nanoparticles by Streptomyces sp. NK52 and its anti-lipid peroxidation activity. *Indian J Exp Biol* [Internet]. 2013 [cited 2025 Aug 8]. Available from: https://www.academia.edu/download/52563987/Biosynthesis_of_colloidal_gold_nanoparti20170409-9057-636sfi.pdf
109. Bukhari SI, Hamed MM, Al-Agamy MH, Gazwi HSS, Radwan HH, Youssif AM. Biosynthesis of copper oxide nanoparticles using streptomyces MHM38 and its biological applications. *J Nanomater* [Internet]. 2021 [cited 2025 Aug 8];2021. Available from: https://onlinelibrary.wiley.com/doi/abs/10.1155/2021/6693302
110. Hassan SE, Fouda A, Radwan AA, Salem SS, Barghoth MG, Awad MA, Abdo AM, El-Gamal MS. Endophytic actinomycetes Streptomyces spp mediated biosynthesis of copper oxide nanoparticles as a promising tool for biotechnological applications. *J Biol Inorg Chem* [Internet]. 2019 May 1 [cited 2025 Aug 8];24(3):377–93. Available from: https://link.springer.com/article/10.1007/s00775-019-01654-5
111. Seth R, Meena A. Enzymes-based nanomaterial synthesis: An eco-friendly and green synthesis approach. *Clean Technol Environ Policy*. 2024 May: 1–24.
112. Mishra A, Sardar M. Alpha-amylase mediated synthesis of silver nanoparticles. *Sci Adv Mater*. 2012 Jan;4(1):143–6.
113. Abada E, Al-Faifi Z, Osman M. *Enzymes and nanoparticles produced by microorganisms and their applications in biotechnology* [Internet]. 2017 [cited 2025 Aug 8];119–50. Available from: https://link.springer.com/chapter/10.1007/978-3-319-68424-6_7
114. Kisailus D, Choi JH, Weaver JC, Yang W, Morse DE. Enzymatic synthesis and nanostructural control of gallium oxide at low temperature. *Adv Mater* [Internet]. 2005 Feb 10 [cited 2025 Aug 8];17(3):314–18. DOI: 10.1002/adma.200400815. Available from: https://onlinelibrary.wiley.com/doi/10.1002/adma.200400815
115. El-Batal AI, Elkenawy NM, Yassin AS, Amin MA. Laccase production by Pleurotus ostreatus and its application in synthesis of gold nanoparticles. *Biotechnol Rep* [Internet]. 2015 Mar 1 [cited 2025 Aug 8];5(1):31–9. Available from: https://www.sciencedirect.com/science/article/pii/S2215017X14000538?via%3Dihub
116. Ovais M, Khalil AT, Ayaz M, Ahmad I, Nethi SK, Mukherjee S. Biosynthesis of metal nanoparticles via microbial enzymes: A mechanistic approach. *Int J Mol Sci* [Internet]. 2018 Dec 18 [cited 2025 Aug 8];19(12):4100. Available from: https://www.mdpi.com/1422-0067/19/12/4100/htm
117. Harne S, Sharma A, Dhaygude M, Joglekar S, Kodam K, Hudlikar M. Novel route for rapid biosynthesis of copper nanoparticles using aqueous extract of Calotropis procera L. latex and their cytotoxicity on tumor cells. *Colloids Surf B Biointerfaces* [Internet]. 2012 Jun 15 [cited 2025 Aug 8];95:284–8. Available from: https://www.sciencedirect.com/science/article/abs/pii/S0927776512001750?via%3Dihub
118. Hudlikar M, Joglekar S, Dhaygude M, Kodam K. Green synthesis of TiO2 nanoparticles by using aqueous extract of Jatropha curcas L. latex. *Mater Lett* [Internet]. 2012 May 15 [cited 2025 Aug 8];75:196–9. Available from: https://www.sciencedirect.com/science/article/abs/pii/S0167577X12002042?via%3Dihub
119. Mukherjee K, Gupta R, Kumar G, Kumari S, Biswas S, Padmanabhan P. Synthesis of silver nanoparticles by Bacillus clausii and computational profiling of nitrate reductase

enzyme involved in production. *J Genet Eng Biotechnol* [Internet]. 2018 Dec 1 [cited 2025 Aug 8];16(2):527–36. Available from: https://www.sciencedirect.com/science/article/pii/S1687157X18300313?via%3Dihub

120. Sharma A, Sharma T, Sharma S, Kumar D, Gondil VS, Mehra N, et al. The impact of nanoparticles-based enzyme immobilization in biocatalysis. *Nanomater Biocatal* [Internet]. 2022 Jan 1 [cited 2025 Aug 8]:149–68. Available from: https://www.sciencedirect.com/science/article/abs/pii/B9780128244364000216?via%3Dihub
121. Ahmad A, Mukherjee P, Mandal D, Senapati S, Khan MI, Kumar R, et al. Enzyme mediated extracellular synthesis of CdS nanoparticles by the fungus, Fusarium oxysporum. *J Am Chem Soc* [Internet]. 2002 Oct 16 [cited 2025 Aug 8];124(41):12108–9. Available from: https://pubs.acs.org/doi/abs/10.1021/ja027296o
122. Kumar SA, Abyaneh MK, Gosavi SW, Kulkarni SK, Pasricha R, Ahmad A, et al. Nitrate reductase-mediated synthesis of silver nanoparticles from AgNO 3. *Biotechnol Lett* [Internet]. 2007 Mar 20 [cited 2025 Aug 8];29(3):439–45. Available from: https://link.springer.com/article/10.1007/s10529-006-9256-7
123. Brayner R, Barberousse H, Hemadi M, Djedjat C, Yéprémian C, Coradin T, et al. Cyanobacteria as bioreactors for the synthesis of Au, Ag, Pd, and Pt nanoparticles via an enzyme-mediated route. *J Nanosci Nanotechnol*. 2007 Aug;7(8):2696–708.
124. Eby DM, Schaeublin NM, Farrington KE, Hussain SM, Johnson GR. Lysozyme catalyzes the formation of antimicrobial silver nanoparticles. *ACS Nano* [Internet]. 2009 Apr 28 [cited 2025 Aug 8];3(4):984–94. Available from: https://pubs.acs.org/doi/abs/10.1021/nn900079e
125. Durán N, Marcato PD, Alves OL, De Souza GIH, Esposito E. Mechanistic aspects of biosynthesis of silver nanoparticles by several Fusarium oxysporum strains. *J Nanobiotechnology* [Internet]. 2005 Jul 13 [cited 2025 Aug 8];3(1):1–7. Available from: https://jnanobiotechnology.biomedcentral.com/articles/10.1186/1477-3155-3-8
126. Shahverdi AR, Minaeian S, Shahverdi HR, Jamalifar H, Nohi AA. Rapid synthesis of silver nanoparticles using culture supernatants of Enterobacteria: A novel biological approach. *Process Biochem* [Internet]. 2007 May 1 [cited 2025 Aug 8];42(5):919–23. Available from: https://www.sciencedirect.com/science/article/abs/pii/S135951130700061X?via%3Dihub
127. Araya-Castro K, Chao TC, Durán-Vinet B, Cisternas C, Ciudad G, Rubilar O. Green synthesis of copper oxide nanoparticles using protein fractions from an aqueous extract of brown algae macrocystis pyrifera. *Processes*. 2021;9(1):1–10.
128. Das SK, Dickinson C, Lafir F, Brougham DF, Marsili E. Synthesis, characterization and catalytic activity of gold nanoparticles biosynthesized with Rhizopus oryzae protein extract. *Green Chem*. 2012;14(5):1322–34.
129. Sanghi R, Verma P. pH dependant fungal proteins in the "green" synthesis of gold nanoparticles. *Adv Mater Lett*. 2010;1(3):193–9.
130. Prabhawathi V, Sivakumar PM, Doble M. Green synthesis of protein stabilized silver nanoparticles using Pseudomonas fluorescens, a marine bacterium, and its biomedical applications when coated on polycaprolactam. *Ind Eng Chem Res*. 2012;51(14):5230–9.
131. Chowdhury S, Basu A, Kundu S. Green synthesis of protein capped silver nanoparticles from phytopathogenic fungus Macrophomina phaseolina (Tassi) Goid with antimicrobial properties against multidrug-resistant bacteria. *Nanoscale Res Lett*. 2014;9(1):1–11.
132. Elshafei AM, Othman AM, Elsayed MA, Al-Balakocy NG, Hassan MM. Green synthesis of silver nanoparticles using Aspergillus oryzae NRRL447 exogenous proteins: Optimization via central composite design, characterization and biological applications. *Environ Nanotechnol, Monit Manage*. 2021 Jul;16.
133. Khamhaengpol A, Siri S. Fluorescent light mediated a green synthesis of silver nanoparticles using the protein extract of weaver ant larvae. *J Photochem Photobiol B Biol* [Internet]. 2016;163:337–44. Available from: http://dx.doi.org/10.1016/j.jphotobiol.2016.09.003

134. Usha Rani P, Rajasekharreddy P. Green synthesis of silver-protein (core-shell) nanoparticles using Piper betle L. leaf extract and its ecotoxicological studies on Daphnia magna. *Colloids Surf A Physicochem Eng Asp* [Internet]. 2011;389(1–3):188–94. Available from: http://dx.doi.org/10.1016/j.colsurfa.2011.08.028
135. Vanga S, Satla SR. A review on green synthesis, characterization and applications of plant mediated metal nanoparticles. *Next Res*. 2025;2(2):100356.
136. Lithi IJ, Ahmed Nakib KI, Chowdhury AMS, Sahadat Hossain M. A review on the green synthesis of metal (Ag, Cu, and Au) and metal oxide (ZnO, MgO, Co3O4, and TiO2) nanoparticles using plant extracts for developing antimicrobial properties. *Nanoscale Adv*. 2025;7(9):2446–73.
137. Aravind Kumar J, Krithiga T, Manigandan S, Sathish S, Annam Renita A, Prakash P, et al. Corrigendum to "a focus to green synthesis of metal/metal based oxide nanoparticles: Various mechanisms and applications towards ecological approach". J Clean Prod. 324 (2021) 129198](S0959652621033849)(10.1016/j.jclepro.2021.129198). *J Clean Prod*. 2021 Apr;328.
138. Bressi V, Ferlazzo A, Iannazzo D, Espro C. Graphene quantum dots by eco-friendly green synthesis for electrochemical sensing: Recent advances and future perspectives. *Nanomaterials* [Internet]. 2021 [cited 2025 Aug 13]. Available from: https://www.mdpi.com/2079-4991/11/5/1120
139. Kalambate PK, Rao Z, Wu J, Shen Y, Boddula R. Biosensors and bioelectronics electrochemical (bio) sensors go green. *Biosens Bioelectron* [Internet]. 2020 May;163:112270. Available from: https://doi.org/10.1016/j.bios.2020.112270
140. Nazarpour S, Hajian R, Sabzvari MH. A novel nanocomposite electrochemical sensor based on green synthesis of reduced graphene oxide/gold nanoparticles modified screen printed electrode for determination of tryptophan using response surface methodology approach. *Microchem J* [Internet]. 2020 [cited 2025 Aug 4]. Available from: https://www.sciencedirect.com/science/article/pii/S0026265X19330735
141. Dayakar T, Rao KV, Bikshalu K, Rajendar V, Park SH. Novel synthesis and structural analysis of zinc oxide nanoparticles for the non enzymatic glucose biosensor. *Mater Sci Eng C* [Internet]. [cited 2025 Aug 4]. Available from: https://www.sciencedirect.com/science/article/pii/S0928493116316642
142. Thirumalraj B, Rajkumar C, Chen SM, Palanisamy S. One-pot green synthesis of graphene nanosheets encapsulated gold nanoparticles for sensitive and selective detection of dopamine. *Sci Rep*. 2017 Jan;7:1–11.
143. Kumar V, Singh K, Panwar S, Mehta SK. Green synthesis of manganese oxide nanoparticles for the electrochemical sensing of p-nitrophenol. *Int Nano Lett*. 2017;7(2):123–31.
144. Palanisamy S, Thirumalraj B, Chen SM, AjmalAli M, Muthupandi K, Emmanuel R, et al. Fabrication of silver nanoparticles decorated on activated screen printed carbon electrode and its application for ultrasensitive detection of dopamine. *Electroanalysis*. 2015;27(8):1998–2006.
145. Sukumar S, Rudrasenan A, Padmanabhan Nambiar D. Green-synthesized rice-shaped copper oxide nanoparticles using caesalpinia bonducella seed extract and their applications. *ACS Omega*. 2020;5(2):1040–51.
146. Mandal N, Mitra R, Trends BPC, 2023 undefined. Bio-synthesized silver nanoparticle modified glassy carbon electrode as electrochemical biosensor for prostate specific antigen detection. *Carbon Trends* [Internet]. 2023 [cited 2025 Aug 5]. Available from: https://www.sciencedirect.com/science/article/pii/S2667056923000706
147. Sharma D, Sabela MI, Kanchi S, Bisetty K, Skelton AA, Honarparvar B. Green synthesis, characterization and electrochemical sensing of silymarin by ZnO nanoparticles: Experimental and DFT studies. *J Electroanal Chem* [Internet]. 2018 Jan 1 [cited 2025 Aug 4];808:160–72. Available from: https://www.sciencedirect.com/science/article/abs/pii/S1572665717308305?via%3Dihub

148. Muthuchamy N, Atchudan R, Edison TNJI, Perumal S, Lee YR. High-performance glucose biosensor based on green synthesized zinc oxide nanoparticle embedded nitrogen-doped carbon sheet. *J Electroanal Chem* [Internet]. 2018 May 1 [cited 2025 Aug 4];816:195–204. Available from: https://www.sciencedirect.com/science/article/abs/pii/S1572665718302340?via%3Dihub
149. Gijare MS, Chaudhari SR, Ekar S, Shaikh SF, Al-Enizi AM, Pandit B, et al. Green synthesis of reduced graphene oxide (rGO) and its applications in non-enzymatic electrochemical glucose sensors. *J Photochem Photobiol A Chem*. 2024;450(Dec 2023).
150. Shalauddin M, Akhter S, Basirun WJ, Akhtaruzzaman M, Mohammed MA, Abd Rahman NM, Salleh NM. Bio-synthesized copper nanoparticle decorated multiwall carbon nanotube-nanocellulose nanocomposite: An electrochemical sensor for the simultaneous detection of acetaminophen and diclofenac sodium. *Surf Interfaces* [Internet]. 2022 [cited 2025 Aug 5]. Available from: https://www.sciencedirect.com/science/article/pii/S2468023022006460
151. Ahmed K, Singh P, Yadav SK, Singh RP, Kumar A, Mishra SK, Singh J. Highly precise electrochemical biosensing of urea through Trigonella foenum-graecum leaf extract mediated green synthesis of titanium dioxide nanoparticles. *Inorg Chem Commun* [Internet]. 2025 [cited 2025 Aug 5]. Available from: https://www.sciencedirect.com/science/article/pii/S1387700325002655
152. Guler M, Meydan I, Seckin H. Electrochemical characterization of vitamin B2 (riboflavin) at Ag nanoparticles synthesized by green chemistry dispersed on reduced graphene oxide. *Diam Relat Mater*. 2023;135.
153. Ettadili FE, Aghris S, Laghrib F, Farahi A, Bakasse M, Lahrich S, El Mhammedi MA. Electrochemical detection of ornidazole in commercial milk and water samples using an electrode based on green synthesis of silver nanoparticles using cellulose separated from Phoenix dactylifera seed. *Int J Biol Macromol* [Internet]. 2023 [cited 2025 Aug 5]. Available from: https://www.sciencedirect.com/science/article/pii/S0141813023018895
154. Somashekar MN, Dhanalakshmi M, Nagamani TS, Chandra TS, Sharanakumar TM, Ravikumar CR. Photocatalytic and electrochemical sensor detection of ascorbic and uric acid using novel plant extract green synthesis of CaO nanoparticles. *Sens Int* [Internet]. 2025 [cited 2025 Aug 5]. Available from: https://www.sciencedirect.com/science/article/pii/S2666351124000305
155. Palan NG, Kiran R, Vinay kumar E, Suma GR, Nagaraju G. Facile one step green synthesis of CdO-CdS hybrid nanocomposite: Its electrochemical and photoluminescence applications. *Nano-Struct Nano-Objects*. 2024;38.
156. Varsha Raj G, Priyadarshini HN, Prashantha K, Nagaraju G, Udayabhanu, Ramakrishnappa T. Green synthesis of Ag-ZnFe2O4@graphene nanocomposite for photocatalytic and electrochemical applications. *Opt Mater*. 2024;147.
157. Avinash B, Ravikumar CR, Basavaraju N, Abebe B, Kumar TN, Manjula SN, et al. Facile green synthesis of zinc oxide nanoparticles: Its photocatalytic and electrochemical sensor for the determination of paracetamol and D-glucose. *Environ Funct Mater* [Internet]. 2023 [cited 2025 Aug 5]. Available from: https://www.sciencedirect.com/science/article/pii/S2773058124000036
158. Siva Ranjani R, Wilson J, Madhu Malar M, Faiyazuddin M, Gowri S. Green synthesized ZnO/NiO nanocomposites decorated tragacanth gum for electrochemical detection of riboflavin. *Microchem J*. 2025;216.
159. Nardi N, Baumgarten LG, Dreyer JP, Santana ER, Winiarski JP, Vieira IC. Nanocomposite based on green synthesis of gold nanoparticles decorated with functionalized multi-walled carbon nanotubes for the electrochemical determination of hydroxychloroquine. *J Pharm Biomed Anal* [Internet]. 2023 [cited 2025 Aug 5]. Available from: https://www.sciencedirect.com/science/article/pii/S0731708523004508

160. Devi BS, Karthikeyan R, Anitha M, Prakash S. Enhanced electrochemical sensing of dopamine using PVP/green synthesized ZnO nanocomposite-modified screen-printed carbon electrode: A promising tool for neurochemical analysis. *Mater Sci Eng B*. 2025 Mar;317.
161. Ettadili FE, Azriouil M, Chhaibi B, Ouatmane FZ, Alaoui OT, Laghrib F, et al. Green synthesis of silver nanoparticles using Phoenix dactylifera seed extract and their electrochemical activity in Ornidazole reduction. *Food Chem Adv* [Internet]. 2023 Oct 1 [cited 2025 Aug 5];2:100146. Available from: https://www.sciencedirect.com/science/article/pii/S2772753X22001344
162. Li H, Li N, Zuo P, Qu S, Qin F, Shen W. Utilization of nitrogen, sulfur co-doped porous carbon micron spheres as bifunctional electrocatalysts for electrochemical detection of cadmium, lead and mercury ions and oxygen evolution reaction. *J Colloid Interface Sci* [Internet]. 2023 [cited 2025 Aug 4]. Available from: https://www.sciencedirect.com/science/article/pii/S0021979723003235
163. Hareesha N, Soumya D, . . . JMJ of, 2024 undefined. Honeycomb polypore biomass-derived activated porous carbon nanosheets/graphite/nafion composite: Green and sensitive electrocatalyst for nanomolar detection of Hg2+ ions and water-splitting reactions. *J Environ Chem Eng* [Internet]. 2024 [cited 2025 Aug 6]. Available from: https://www.sciencedirect.com/science/article/pii/S2213343724017147
164. Djebbi MA, Allagui L, El Ayachi MS, Boubakri S, Jaffrezic-Renault N, Namour P, et al. Zero-valent iron nanoparticles supported on biomass-derived porous carbon for simultaneous detection of Cd2+ and Pb2+. *ACS Appl Nano Mater* [Internet]. 2022 Jan 28 [cited 2025 Aug 4];5(1):546–58. Available from: https://pubs.acs.org/doi/abs/10.1021/acsanm.1c03333
165. Guan J, Fang Y, Zhang T, Wang L, Zhu H, Du M, et al. Kelp-derived activated porous carbon for the detection of heavy metal ions via square wave anodic stripping voltammetry. *Electrocatalysis* [Internet]. 2020 Jan 1 [cited 2025 Aug 4];11(1):59–67. Available from: https://link.springer.com/article/10.1007/s12678-019-00568-9
166. El Hamdouni Y, El Hajjaji S, Szabo T, Trif L, Felhősi I, Abbi K, et al. Biomass valorization of walnut shell into biochar as a resource for electrochemical simultaneous detection of heavy metal ions in water and soil samples. *Arab J Chem* [Internet]. 2022 [cited 2025 Aug 4]. Available from: https://www.sciencedirect.com/science/article/pii/S1878535222005688
167. Yin M, Pu H, Zhang Y, Gao P, Sun Q, Yin W, et al. Electrochemical sensor based on biomass yeast integrated sulfur-doped graphene and carboxylated carbon nanotubes/MoS2 for highly-sensitive detection of Pb2+. *Electroanalysis* [Internet]. 2023 Feb 1 [cited 2025 Aug 4];35(2). Available from: https://analyticalsciencejournals.onlinelibrary.wiley.com/doi/abs/10.1002/elan.202200163?casa_token=nvLEGG0NlW4AAAAA:EhSb7Ip-JUS93g5zcN1vF-uxUXt-0B5hIvSIf5w6qZ0TGjraYrR4toUzuBsCs3NA2X7VTM1Qp-BKude7C5
168. Sanni SO, Moundzounga TH, Oseghe EO, Haneklaus NH, Viljoen EL, Brink HG. One-step green synthesis of water-soluble fluorescent carbon dots and its application in the detection of Cu2+. *Nanomaterials* [Internet]. 2022 [cited 2025 Aug 5]. Available from: https://www.mdpi.com/2079-4991/12/6/958?trk=public_post_main-feed-card-text
169. Zou J, Liu J, Yu Q, Gao Y, Chen S, Huang X, et al. Facile synthesis of nitrogen self-doped porous carbon derived from cicada shell via KOH activation for simultaneous detection and removal of Cu2+. *Mol* [Internet]. 2022 [cited 2025 Aug 5]. Available from: https://www.mdpi.com/1420-3049/27/14/4516
170. Zhu X, Liu B, Chen S, Wu L, Yang J, Liang S, et al. Ultrasensitive and simultaneous electrochemical determination of Pb2+ and Cd2+ based on biomass derived lotus root-like hierarchical porous carbon/bismuth composite. *J Electrochem Soc* [Internet]. 2020 [cited

2025 Aug 5]. Available from: https://iopscience.iop.org/article/10.1149/1945-7111/ab82f7/meta

171. Baikeli Y, Mamat X, Yalikun N, Wang Y, Qiao M, Li Y, et al. Differential pulse voltammetry detection of Pb (ii) using nitrogen-doped activated nanoporous carbon from almond shells. *RSC Adv* [Internet]. 2019 [cited 2025 Aug 5]. Available from: https://pubs.rsc.org/en/content/articlehtml/2019/ra/c9ra03925b
172. Boobalan T, Sethupathi M, Sengottuvelan N, Kumar P, Balaji P, Gulyás B, et al. Mushroom-derived carbon dots for toxic metal ion detection and as antibacterial and anticancer agents. *ACS Appl Nano Mater* [Internet]. 2020 Jun 26 [cited 2025 Aug 5];3(6):5910–19. Available from: https://pubs.acs.org/doi/abs/10.1021/acsanm.0c01058
173. Zeinu KM, Hou H, Liu B, Yuan X, Huang L, Zhu X, et al. A novel hollow sphere bismuth oxide doped mesoporous carbon nanocomposite material derived from sustainable biomass for picomolar electrochemical detection of lead and cadmium. *J Mater Chem A* [Internet]. 2016 [cited 2025 Aug 5]. Available from: https://pubs.rsc.org/en/content/articlehtml/2016/ta/c6ta04881a
174. Manoj D, Saravanan R, Santhanalakshmi J, Agarwal S, Gupta VK, Boukherroub R. Towards green synthesis of monodisperse Cu nanoparticles: An efficient and high sensitive electrochemical nitrite sensor. *Sens Actuators B Chem* [Internet]. 2018 [cited 2025 Aug 4]. Available from: https://www.sciencedirect.com/science/article/pii/S0925400518306397
175. Salazar P, Fernández I, Rodríguez MC, Hernández-Creus A, González-Mora JL. One-step green synthesis of silver nanoparticle-modified reduced graphene oxide nanocomposite for H2O2 sensing applications. *J Electroanal Chem*. 2019;855:113638.
176. Vatandost E, Ghorbani-HasanSaraei A, Chekin F, Naghizadeh Raeisi S, Shahidi SA. Green tea extract assisted green synthesis of reduced graphene oxide: Application for highly sensitive electrochemical detection of sunset yellow in food products. *Food Chem X* [Internet]. 2020 Jun 30 [cited 2025 Aug 4];6:100085. Available from: https://www.sciencedirect.com/science/article/pii/S2590157520300092?via%3Dihub
177. Ikhsan NI, Rameshkumar P, Pandikumar A, Shahid MM, Huang NM, Kumar SV, et al. Facile synthesis of graphene oxide–silver nanocomposite and its modified electrode for enhanced electrochemical detection of nitrite ions. *Talanta* [Internet]. 2015 [cited 2025 Aug 4]. Available from: https://www.sciencedirect.com/science/article/pii/S0039914015301788
178. Palanisamy S, Karuppiah C, Chen SM, Muthupandi K, Emmanuel R, Prakash P, et al. Selective and simultaneous determination of dihydroxybenzene isomers based on green synthesized gold nanoparticles decorated reduced graphene oxide. *Electroanalysis*. 2015;27(5):1144–51.
179. Wicaksono WP, Fadilla NI, Zamar AA, Fadillah G, Anugrahwati M, Anas AK, et al. Formaldehyde electrochemical sensor using graphite paste-modified green synthesized zinc oxide nanoparticles. *Inorg Chem Commun* [Internet]. 2022 Sep 1 [cited 2025 Aug 4];143:109729. Available from: https://www.sciencedirect.com/science/article/abs/pii/S1387700322005378?via%3Dihub
180. Turunc E, Kahraman O, Binzet R. Green synthesis of silver nanoparticles using pollen extract: Characterization, assessment of their electrochemical and antioxidant activities. *Anal Biochem* [Internet]. 2021 May 15 [cited 2025 Aug 4];621:114123. Available from: https://www.sciencedirect.com/science/article/abs/pii/S0003269721000245?via%3Dihub
181. Bodur OC, Keskin M, Keskin Ş, Arslan F. Green synthesis of Ananas comosus (L.) Merr. based-silver nanoparticles and determination of their potential in electrochemical detection of H2O2. *Biomass Convers Biorefinery*. 2025:0123456789.
182. Arularasu MV, Rajendran TV, Vignesh R, Nelson VK, Yusuf SM. A screen-printed carbon electrode modified with a green synthesized gold nanoparticle for selective

detection of nitrite in drinking water sample. *Sens Biosensing Res* [Internet]. 2025 [cited 2025 Aug 5]. Available from: https://www.sciencedirect.com/science/article/pii/S2214180425000479

183. Murugan S, Srinivasan G, Kotteeswaran S, Jeyaraj W, Govindasamy M, Thangavel K. Green synthesis of Ag2O/ZnO nanocomposite for efficient antibacterial property and tunable electrochemical detection of hydrogen peroxide. *J Environ Chem Eng*. 2025;13.
184. Mane SA, Moholkar AV, Ghule AV. A green approach for synthesis of an efficient nano pebbles-like BiVO4@C electrode for supercapacitor application. *Next Mater*. 2023;1.
185. Thalthodi MS, Sahadevan J, Sasikumar R, Muthu SE, Padmanathan N. Bio-extract mediated green synthesis of bi-functional ammonium cobalt phosphate nanoflakes electrode/catalyst for supercapattery and ethanol oxidation. *Inorg Chem Commun*. 2024;164.
186. Lavanya V, Santhakumar K. A study on bio-synthesized gold nanoparticles via Hibiscus Platanifolius flower extract: Efficient strategy for cytotoxicity, photocatalytic potential, and green supercapacitor applications. *Inorg Chem Commun* [Internet]. 2025 [cited 2025 Aug 7]. Available from: https://www.sciencedirect.com/science/article/pii/S1387700324018768
187. Sobi MA, Bindhu MR, Anjana PM, Usha D, Rajagopal R, Alfarhan A, et al. Green synthesis of Nyctanthes arbor-tristis flower-decorated gold nanoparticles: Sustainable approaches for enhancing antimicrobial and supercapacitor performance. *Process Saf Environ Prot*. 2024;187:59–72.
188. Arjunkumar B, Mangalaraja RV, Ragupathi C, Ramalingam G, Alam MM, Al-Sehemi AG, et al. Green carbon nanospheres for integrated photo-supercapacitors device with improved efficiency. *Diam Relat Mater*. 2024 Oct;150.
189. Harini HV, Nagaswarupa HP, Naik R, Ravikumar CR. Green mediated sol-gel auto combustion and solution combustion synthesis of calcium magnesium aluminate (CMA) nanoparticles: Modified carbon paste and nickel mesh electrodes for supercapacitor and sensor applications. *J Indian Chem Soc*. 2023;100.
190. Rohit M, Basavannavar DR, Praveen BM, Kumar N, Rajana GS, et al. Green synsthesis of copper molybedate and its various application in sensing, supercapacitors and photocatalytic activity. *Nano Trends*. 2025;10.
191. Ugwuoke CO, Ghali M, El-Moneim AA. Green synthesis of carbon dots from Nigella sativa seeds for supercapacitor application. *J Energy Storage*. 2024;95.
192. Başlak C, Öztürk G, Demirel S, Kocyigit A, Doğu S, Yıldırım M. Green synthesis of carbon quantum dots from Sideritis vuralii and its application in supercapacitors. *Inorg Chem Commun*. 2023;153.
193. Elattar MS, Emara MM, Ali AE, Kashyout AE. High-yield green synthesis of carbon-nanohair/carbon quantum dots/activated carbon composite from agricultural or textile wastes for enhanced supercapacitor performance. *J Phys Chem Solids* [Internet]. 2025 [cited 2025 Aug 7]. Available from: https://www.sciencedirect.com/science/article/pii/S0022369725003476
194. Karthikeyan M, Rajkumar S, Priyadharshan M. Green synthesis of Ce doped ZnO nanoparticles using phyllanthus emblica seed extract for the efficient high-performance supercapacitor application. *J Ind Eng Chem*. 2025;147:608–18.
195. Ganesh CUC, Krushna BRR, Pruthviraj IS, Ramakrishna G, Sharma SC, George A, et al. Green synthesis of Ce3+ doped V2O5 NPs as an advanced electrode material for possible supercapacitor and therapeutic applications. *J Taiwan Inst Chem Eng*. 2025;174.
196. Iskender HY, Eksik O, Arvas MB, Yazar S. Green Synthesis of MXene@ TiO_2 Nanocomposites for high-performance flexible supercapacitors: Synergistic enhancement via polypyrrole polymerization and dopant engineering. *Synth Met* [Internet]. 2025 [cited 2025 Aug 7]. Available from: https://www.sciencedirect.com/science/article/pii/S0379677925000918

197. Prabhu R, Jayarambabu N, Anitha N, Jitesh P, Hitesh B, Venkatappa Rao T. Size dependent electrochemical properties of green synthesized NiO nanoparticles as a supercapacitor electrode. *Inorg Chem Commun*. 2024;160(Dec 2023).
198. Gunasekaran SS, Gopalakrishnan A, Subashchandrabose R, Badhulika S. Phytogenic generation of NiO nanoparticles as green-electrode material for high performance asymmetric supercapacitor applications. *J Energy Storage* [Internet]. 2021 [cited 2025 Aug 7]. Available from: https://www.sciencedirect.com/science/article/pii/S2352152X21001663
199. Nwanya AC, Ndipingwi MM, Ikpo CO, Obodo RM, Nwanya SC, Botha S, et al. Zea mays lea silk extract mediated synthesis of nickel oxide nanoparticles as positive electrode material for asymmetric supercabattery. *J Alloy Compd* [Internet]. 2020 [cited 2025 Aug 7]. Available from: https://www.sciencedirect.com/science/article/pii/S0925838819348273
200. Sone BT, Fuku XG, Maaza M. Physical & electrochemical properties of green synthesized bunsenite NiO nanoparticles via Callistemon viminalis' extracts. *Int J Electrochem Sci* [Internet]. 2016 [cited 2025 Aug 7]. Available from: https://www.sciencedirect.com/science/article/pii/S145239812317725X
201. Kundu M, Karunakaran G, Kuznetsov D. Green synthesis of NiO nanostructured materials using Hydrangea paniculata flower extracts and their efficient application as supercapacitor electrodes. *Powder Technol* [Internet]. 2017 [cited 2025 Aug 7]. Available from: https://www.sciencedirect.com/science/article/pii/S0032591017301043
202. Karpagavinayagam P, Prasanna AE, Vedhi C. Eco-friendly synthesis of nickel oxide nanoparticles using Avicennia Marina leaf extract: Morphological characterization and electrochemical application. *Mater Today Proc* [Internet]. 2022 [cited 2025 Aug 7]. Available from: https://www.sciencedirect.com/science/article/pii/S221478532032798X
203. Nivedhitha Bharathi A, Maheshwaran G, Kousalya Devi V, Krishna Kumar M, Sudhahar S. Enhanced electrochemical performance of bio-derived nickel oxide nanoparticles using Zephyranthes rosea flower extract and investigation on charge storage. *Bull Mater Sci* [Internet]. 2022 Dec 1 [cited 2025 Aug 7];45(4). Available from: https://link.springer.com/article/10.1007/s12034-022-02773-7
204. Lamba P, Singh P, Singh P, Kumar A, Singh P, Bharti, et al. Bioinspired synthesis of nickel oxide nanoparticles as electrode material for supercapacitor applications. *Ionics* [Internet]. 2021 Dec 1 [cited 2025 Aug 7];27(12):5263–76. Available from: https://link.springer.com/article/10.1007/s11581-021-04245-0
205. Atalay FE, Asma D, Kaya H, Bingol A, Yaya P. Synthesis of NiO nanostructures using cladosporium cladosporioides fungi for energy storage applications. *Nanomater Nanotechnol* [Internet]. 2016 [cited 2025 Aug 7];6(1). Available from: https://journals.sagepub.com/doi/abs/10.5772/63569
206. Gebretinsae HG, Welegergs GG, Maaza M, Gebretinsae H, Welegergs G, Matinise N, et al. Electrochemical study of Nickel Oxide (NiO) nanoparticles from cactus plant extract. *MRS Adv* [Internet]. 2020 [cited 2025 Aug 7]. Available from: https://www.cambridge.org/core/journals/mrs-advances/article/electrochemical-study-of-nickel-oxide-nio-nanoparticles-from-cactus-plant-extract/2131D264EE70DBC8DD89AE2EF3E39B93
207. Sumantha HS, Suresha BL. Green, cost-effective synthesis of NiMnO3 nanoparticles and their use in supercapacitor and photodegradation applications. *Nano-Struct Nano-Objects*. 2024;38.
208. Dhanemozhi AC, Rajeswari V, Sathyajothi S. Green synthesis of zinc oxide nanoparticle using green tea leaf extract for supercapacitor application. *Mater Today Proc*. 2017;4:660–7.
209. Yadav S, Rani N, Saini K. Green synthesis of ZnO and CuO NPs using Ficus benghalensis leaf extract and their comparative study for electrode materials for high performance supercapacitor application. *Mater Today Proc* [Internet]. 2022 Jan 1 [cited 2025 Aug

7];49:2124–30. Available from: https://www.sciencedirect.com/science/article/abs/pii/S2214785321057680?via%3Dihub
210. Rani N, Saini M, Yadav S, Gupta K, Saini K, Khanuja M. High performance super-capacitor based on rod shaped ZnO nanostructure electrode. *AIP Conf Proc* [Internet]. 2020 Oct 29 [cited 2025 Aug 7];2276(1). Available from: https://pubs.aip.org/aip/acp/article-abstract/2276/1/020042/777380/High-performance-super-capacitor-based-on-rod?redirectedFrom=fulltext
211. Sasirekha C, Arumugam S, Muralidharan G. Green synthesis of ZnO/carbon (ZnO/C) as an electrode material for symmetric supercapacitor devices. *Applied Surf Sci* [Internet]. 2018 Aug 15 [cited 2025 Aug 7];449:521–7. Available from: https://www.sciencedirect.com/science/article/pii/S0169433218301843
212. Kalaiarasi J, Pragathiswaran C, Subramani P. Green chemistry approach for the functionalization of reduced graphene and ZnO as efficient supercapacitor application. *J Mol Struct* [Internet]. 2021 Oct 15 [cited 2025 Aug 7];1242:130704. Available from: https://www.sciencedirect.com/science/article/abs/pii/S0022286021008371
213. Shaheen I, Ahmad KS, Zequine C, Gupta RK, Thomas AG, Malik MA. Facile ZnO-based nanomaterial and its fabrication as a supercapacitor electrode: Synthesis, characterization and electrochemical studies. *RSC Adv* [Internet]. 2021 Jul 2 [cited 2025 Aug 7];11(38):23374–84. Available from: https://pubs.rsc.org/en/content/articlehtml/2021/ra/d1ra04341b
214. Shaheen I, Ahmad KS, Zequine C, Gupta RK, Thomas AG, Malik MA. Green synthesis of ZnO–Co 3 O 4 nanocomposite using facile foliar fuel and investigation of its electrochemical behaviour for supercapacitors. *New J Chem* [Internet]. 2020 Nov 14 [cited 2025 Aug 7];44(42):18281–92. Available from: https://pubs.rsc.org/en/content/articlehtml/2020/nj/d0nj03430d
215. Chakraborty S, Mary NL. Biocompatible supercapacitor electrodes using green synthesised ZnO/Polymer nanocomposites for efficient energy storage applications. *J Energy Storage* [Internet]. 2020 [cited 2025 Aug 7]. Available from: https://www.sciencedirect.com/science/article/pii/S2352152X19313921
216. Choubari MS, Mazloom J, Ghodsi FE. Supercapacitive properties, optical band gap, and photoluminescence of CeO2-ZnO nanocomposites prepared by eco-friendly green and citrate sol-gel methods: a comparative study. *Ceramics Int* [Internet]. 2022 Aug 1 [cited 2025 Aug 7];48(15):21385–95. Available from: https://www.sciencedirect.com/science/article/pii/S0272884222012664
217. Nayak S, Kittur AA, Nayak S. Biosynthesis of zinc oxide-cobalt oxide nanocomposite as electrode material and its performance evaluation for the sustainable hybrid supercapacitor energy storage devices. *Chem Phys Lett* [Internet]. 2022 Nov 1 [cited 2025 Aug 7];806:140058. Available from: https://www.sciencedirect.com/science/article/abs/pii/S0009261422007151
218. Ikhioya IL, Onoh EU, Nkele AC, Abor BC, Obitte BC, Maaza M, Ezema FI. The the green synthesis of copper oxide nanoparticles using the Moringa oleifera plant and its subsequent characterization for use in energy storage applications. *East Eur J Physics* [Internet]. 2023 [cited 2025 Aug 7]. Available from: https://periodicals.karazin.ua/eejp/article/view/21046
219. Ravichandran S, Radhakrishnan J, Sengodan P, Rajendran R. Biosynthesis of copper oxide nanoparticle from clerodendrum phlomidis and their decoration with graphene oxide for photocatalytic and supercapacitor application. *J Mater Sci Mater Electron* [Internet]. 2022 Apr 1 [cited 2025 Aug 7];33(12):9403–11. Available from: https://link.springer.com/article/10.1007/s10854-021-07340-0
220. Sudhakar YN, Hemant H, Nitinkumar SS, Poornesh P, Selvakumar M. Green synthesis and electrochemical characterization of rGO-CuO nanocomposites for supercapacitor applications. *Ionics* [Internet]. 2017 May 14 [cited 2025 Aug 7];23(5):1267–76. Available from: https://link.springer.com/article/10.1007/s11581-016-1923-7

221. Nsude HE, Nsude KU, Whyte GM, Obodo RM, Iroegbu C, Maaza M, et al. Green synthesis of CuFeS2 nanoparticles using mimosa leaves extract for photocatalysis and supercapacitor applications. *J Nanoparticle Res* [Internet]. 2020 Nov 1 [cited 2025 Aug 7];22(11). Available from: https://link.springer.com/article/10.1007/s11051-020-05071-7
222. Alturki AM. Benign feature for copper oxide nanoparticle synthesis using sugarcane molasses and its applications in electrical conductivity and supercapacitor. *Biomass Conv Biorefin* [Internet]. 2024 Jun 1 [cited 2025 Aug 7];14(12):12981–92. Available from: https://link.springer.com/article/10.1007/s13399-022-03303-5
223. Nwanya AC, Ndipingwi MM, Mayedwaa N, Razanamahandry LC, Ikpo CO, Waryo T, et al. Maize (Zea mays L.) fresh husk mediated biosynthesis of copper oxides: Potentials for pseudo capacitive energy storage. *Electrochimica Acta* [Internet]. 2019 Apr 1 [cited 2025 Aug 7];301:436–48. Available from: https://www.sciencedirect.com/science/article/pii/S0013468619302208
224. Kumar JS, Jana M, Khanra P, Samanta P, Koo H, Murmu NC, et al. One pot synthesis of Cu2O/RGO composite using mango bark extract and exploration of its electrochemical properties. *Electrochimica Acta* [Internet]. 2016 Mar 1 [cited 2025 Aug 7];193:104–15. Available from: https://www.sciencedirect.com/science/article/pii/S0013468616303024
225. Ismail E, Khamlich S, Dhlamini M, Maaza M. Green biosynthesis of ruthenium oxide nanoparticles on nickel foam as electrode material for supercapacitor applications. *RSC Adv* [Internet]. 2016 [cited 2025 Aug 7];6(90):86843–50. Available from: https://pubs.rsc.org/en/content/articlehtml/2016/ra/c6ra17996g
226. Nisha B, Vidyalakshmi Y. Enhanced formation of ruthenium oxide nanoparticles through green synthesis for highly efficient supercapacitor applications. *Adv Powder Technol* [Internet]. 2020 [cited 2025 Aug 7]. Available from: https://www.sciencedirect.com/science/article/pii/S0921883119304728
227. Naresh Kumar Reddy P, Shaik DPMD, Ganesh V, Nagamalleswari D, Thyagarajan K, Vishnu Prasanth P. High electrochemical activity of 3D flower like nanostructured TiO2 obtained by green synthesis. *Applied Surf Sci* [Internet]. 2021 Sep 30 [cited 2025 Aug 7];561. Available from: https://www.sciencedirect.com/science/article/pii/S0169433221011685
228. Waris, Hakeem Anwer A, Abdulaziz F, Latif S, Alanazi A, Sultana S, et al. Microwave assisted green synthesis of high capacitive TiO2 doped rGO nanosheets for supercapacitor applications. *Mater Sci Eng B*. 2023;291.
229. Kumar R, Gokul S, Ran F, Sambasivam S, Alrashidi KA, Thangappan R. Green synthesis of g-C3N4 decorated with SnO2 nanocomposites using a novel Ananas comosus crown extract for boosted performance of asymmetric supercapacitors. *J Energy Storage* [Internet]. 2024 [cited 2025 Aug 7]. Available from: https://www.sciencedirect.com/science/article/pii/S2352152X24028172
230. Wang Y, Chang H, Ma T, Deng H, Zha Z. Effect of cotton stalk particle size on the structure of biochar and the performance of anode for lithium-ion battery. *J Phys Chem Solids* [Internet]. 2022 [cited 2025 Aug 8]. Available from: https://www.sciencedirect.com/science/article/pii/S0022369722002736
231. Pan P, Hu Y, Wu K, Cheng Z, Shen Z, Jiang L, et al. Growth of ZnCo2O4 nanocubes on flexible biochar substrate derived from natural silk waste fabric for lithium-ion battery anode. *J Alloy Compd* [Internet]. 2020 Jan 25 [cited 2025 Aug 8];814. Available from: https://www.sciencedirect.com/science/article/pii/S0925838819335522
232. Simões Dos Reis G, Mayandi Subramaniyam C, Cárdenas AD, Larsson SH, Thyrel M, Lassi U, et al. Facile synthesis of sustainable activated biochars with different pore structures as efficient additive-carbon-free anodes for lithium-and sodium-ion batteries. *ACS Omega* [Internet]. 2022 Nov 22 [cited 2025 Aug 8];7(46):42570–81. Available from: https://pubs.acs.org/doi/abs/10.1021/acsomega.2c06054

233. Sun C, Pan J, Zhao X, Jiao C, Yao W, Wang C, et al. Effective coating of Si@ NiO nanoflowers with nitrogen-doped wheat protein-derived biochar for efficient lithium-ion and lithium-sulfur batteries anode materials. *J Alloy Compd* [Internet]. 2023 Dec 15 [cited 2025 Aug 8];968. Available from: https://www.sciencedirect.com/science/article/pii/S0925838823032255
234. Sang J, Sun C, Pan J, Gao C, Zhang R, Jia F, et al. Seaweed – modification of si by natural nitrogen-doped porous biochar for high-efficiency lithium batteries. *ACS Appl Mater Interfaces* [Internet]. 2024 Mar 6 [cited 2025 Aug 8];16(9):11389–99. Available from: https://pubs.acs.org/doi/abs/10.1021/acsami.3c15459
235. Sun C, Du A, Deng G, Zhao X, Pan J, Fu X, et al. Naturally nitrogen-doped self-encapsulated biochar materials based on mouldy wheat flour for silicon anode in lithium-ion batteries. *Electrochimica Acta* [Internet]. 2023 May 10 [cited 2025 Aug 8];450. Available from: https://www.sciencedirect.com/science/article/pii/S0013468623004498
236. Nie S, Li R, Xin Y, Tan Y, Miao C, Xiang Y, et al. Enhanced cycling performance of Sn nanoparticles embedded into the pyrolytic biochar from tea-seed shells as composite anode materials for lithium ions batteries. *Solid State Ionics* [Internet]. 2021 Oct 1 [cited 2025 Aug 8];368. Available from: https://www.sciencedirect.com/science/article/pii/S0167273821001569
237. Zhu L, Zhang W, Chen J, Men L, Zhang J, Zhou Y. Deciphering the storage mechanism of biochar anchored with different morphology Mn3O4 as advanced anode material for lithium-ion batteries. *J Colloid Interface Sci* [Internet]. 2024 Sep 1 [cited 2025 Aug 8];669:740–53. Available from: https://www.sciencedirect.com/science/article/pii/S0021979724010154
238. Zhang X, Wang Y, Ju N, Ai Y, Liu Y, Liang J, et al. Ultimate resourcization of waste: Crab shell-derived biochar for antimony removal and sequential utilization as an anode for a Li-ion battery. *ACS Sustain Chem Eng* [Internet]. 2021 Jul 5 [cited 2025 Aug 8];9(26):8813–23. Available from: https://pubs.acs.org/doi/abs/10.1021/acssuschemeng.1c01851
239. Zhang J, Tahmasebi A, Omoriyekomwan JE, Yu J. Microwave-assisted synthesis of biochar-carbon-nanotube-NiO composite as high-performance anode materials for lithium-ion batteries. *Fuel Process Technol* [Internet]. 2021 Mar 1 [cited 2025 Aug 8];213. Available from: https://www.sciencedirect.com/science/article/pii/S0378382020310055
240. Zhou X, Chen F, Bai T, Long B, Liao Q, Ren Y, et al. Interconnected highly graphitic carbon nanosheets derived from wheat stalk as high performance anode materials for lithium ion batteries. *Green Chem* [Internet]. 2016 [cited 2025 Aug 8];18(7):2078–88. Available from: https://pubs.rsc.org/en/content/articlehtml/2016/gc/c5gc02122g
241. Ou J, Yang L, Xi X. Biomass inspired nitrogen doped porous carbon anode with high performance for lithium ion batteries. *Chinese J Chem* [Internet]. 2016 Jul 1 [cited 2025 Aug 8];34(7):727–32. Available from: https://onlinelibrary.wiley.com/doi/abs/10.1002/cjoc.201600095?casa_token=vVDL67AJq_YAAAAA:SPnpc0g_LFAnbMjjyGCvNjd-plNpkxJ1mI2Vdn8z9oho2n1f1xSwJ4rdXObx_ZwdQzNnW0JNxTuq-IzA
242. Choi C, Seo SD, Kim BK, Kim DW. Enhanced lithium storage in hierarchically porous carbon derived from waste tea leaves. *Sci Rep* [Internet]. 2016 Dec 14 [cited 2025 Aug 8];6. Available from: https://www.nature.com/articles/srep39099
243. Ru H, Xiang K, Zhou W, Zhu Y, Song X, Chen ZH, et al. Bean-dreg-derived carbon materials used as superior anode material for lithium-ion batteries. *Electrochimica Acta* [Internet]. 2016 [cited 2025 Aug 8]. Available from: https://www.sciencedirect.com/science/article/pii/S001346861632285X
244. Zhang D, Wang G, Xu L, Lian J, Bao J, Zhao Y, et al. Defect-rich N-doped porous carbon derived from soybean for high rate lithium-ion batteries. *Appl Surf Sci* [Internet]. 2018 [cited 2025 Aug 8]. Available from: https://www.sciencedirect.com/science/article/pii/S0169433218312455

245. Zheng P, Liu T, Zhang J, Zhang L, Liu Y, Huang J, et al. Sweet potato-derived carbon nanoparticles as anode for lithium ion battery. *RSC Adv* [Internet]. 2015 [cited 2025 Aug 8];5(51):40737–41. Available from: https://pubs.rsc.org/en/content/articlehtml/2015/ra/c5ra03482e
246. Drews M, Büttner J, Bauer M, Ahmed J, Sahu R, Scheu C, et al. Spruce hard carbon anodes for lithium-ion batteries. *ChemElectroChem* [Internet]. 2021 Dec 13 [cited 2025 Aug 8];8(24):4750–61. Available from: https://chemistry-europe.onlinelibrary.wiley.com/doi/abs/10.1002/celc.202101174
247. Zhang W, Li M, Chen Y, Zhang Z. Facile synthesis of honeycomb-structured biochar with high specific surface and its electrochemical properties as anodes for Na-ion batteries. *Ionics* [Internet]. 2023 Dec 4 [cited 2025 Aug 8];29(12):5195–203. Available from: https://link.springer.com/article/10.1007/s11581-023-05235-0
248. Yu K, Wang X, Yang H, Bai Y, Wu C. Insight to defects regulation on sugarcane waste-derived hard carbon anode for sodium-ion batteries. *J Energy Chem* [Internet]. 2021 [cited 2025 Aug 8]. Available from: https://www.sciencedirect.com/science/article/pii/S2095495620305209
249. Yang L, Hu M, Zhang H, Yang W, Lv R. Pore structure regulation of hard carbon: Towards fast and high-capacity sodium-ion storage. *J Colloid Interface Sci* [Internet]. 2020 [cited 2025 Aug 8]. Available from: https://www.sciencedirect.com/science/article/pii/S0021979720300990
250. Jin Q, Li W, Wang K, Li H, Feng P, Zhang Z, et al. Tailoring 2D heteroatom-doped carbon nanosheets with dominated pseudocapacitive behaviors enabling fast and high-performance sodium storage. *Adv Funct Mater* [Internet]. 2020 Apr 1 [cited 2025 Aug 8];30(14). Available from: https://advanced.onlinelibrary.wiley.com/doi/abs/10.1002/adfm.201909907?casa_token=LwTbKUPLhgMAAAAA:lGPdTOYzZ8O9X3M_mT4825BWIvAkv5sOysl40Mv8U5_bSKjmsPZ4YOVwBDFjQl8-TWW-VI1Qfpr-78g
251. Yu Y, Ren Z, Shang Q, Han J, Li L, Chen J, Fakudze S, Tian Z, Liu C. Ionic liquid-induced low temperature graphitization of cellulose-derived biochar for high performance sodium storage. *Surf Coatings Technol* [Internet]. 2021 [cited 2025 Aug 8]. Available from: https://www.sciencedirect.com/science/article/pii/S0257897221002073
252. Khan M, Ahmad N, Lu K, Sun Z, Wei C, Zheng X, et al. Nitrogen-doped carbon derived from onion waste as anode material for high performance sodium-ion battery. *Solid State Ionics* [Internet]. 2020 [cited 2025 Aug 8]. Available from: https://www.sciencedirect.com/science/article/pii/S016727381931207X
253. Dan R, Chen W, Xiao Z, Li P, Liu M, Chen Z, et al. N-doped biomass carbon/reduced graphene oxide as a high-performance anode for sodium-ion batteries. *Energy Fuels* [Internet]. 2020 [cited 2025 Aug 8]. Available from: https://pubs.acs.org/doi/abs/10.1021/acs.energyfuels.0c00058
254. Chen L, Bai L, Yeo J, Wei T, Chen W, Fan Z. Wood-derived carbon with selectively introduced C= O groups toward stable and high capacity anodes for sodium storage. *ACS Appl Mater Interfaces* [Internet]. 2020 Jun 17 [cited 2025 Aug 8];12(24):27499–507. Available from: https://pubs.acs.org/doi/abs/10.1021/acsami.0c04469
255. Alvin S, Yoon D, Chandra C, Cahyadi HS, Park JH, Chang W, et al. Revealing sodium ion storage mechanism in hard carbon. *Carbon* [Internet]. 2019 [cited 2025 Aug 8]. Available from: https://www.sciencedirect.com/science/article/pii/S0008622318312521
256. Nita C, Zhang B, Dentzer J, Ghimbeu CM. Hard carbon derived from coconut shells, walnut shells, and corn silk biomass waste exhibiting high capacity for Na-ion batteries. *J Energy Chem* [Internet]. 2021 [cited 2025 Aug 8]. Available from: https://www.sciencedirect.com/science/article/pii/S2095495620306306
257. Panda MR, Dutta DP, Mitra S. Bio-derived mesoporous disordered carbon: An excellent anode in sodium-ion battery and full-cell lab prototype. *Carbon* [Internet]. 2019

[cited 2025 Aug 8]. Available from: https://www.sciencedirect.com/science/article/pii/S0008622318310698

258. Zhang Y, Li X, Dong P, Wu G, Xiao J, Zeng X, et al. Honeycomb-like hard carbon derived from pine pollen as high-performance anode material for sodium-ion batteries. *ACS Appl Mater Interfaces* [Internet]. 2018 Dec 12 [cited 2025 Aug 8];10(49):42796–803. Available from: https://pubs.acs.org/doi/abs/10.1021/acsami.8b13160
259. Wang P, Zhu K, Ye K, Gong Z, Liu R, Cheng K, et al. Three-dimensional biomass derived hard carbon with reconstructed surface as a free-standing anode for sodium-ion batteries. *J Colloid Interface Sci* [Internet]. 2020 [cited 2025 Aug 8]. Available from: https://www.sciencedirect.com/science/article/pii/S0021979719314195
260. Ahmad A, Muzammal S, Khalid H, Tariq S, Sajid A, Bocchetta P. Biomass-based ORR and OER electrocatalysts: Methods of synthesis and cell performance. *Electrochimica Acta* [Internet]. 2025 [cited 2025 Aug 13]. Available from: https://www.sciencedirect.com/science/article/pii/S0013468625010989
261. Jamesh MI, Akila A, Sudha D, Gnana Priya K, Sivaprakash V, Revathi A. Fabrication of earth-abundant electrocatalysts based on green-chemistry approaches to achieve efficient alkaline water splitting – a review. *Sustainability (Switzerland)*. 2022;14.
262. Su C, Liu Y, Luo Z, Veder JP, Zhong Y, Jiang SP, Shao Z. Defects-rich porous carbon microspheres as green electrocatalysts for efficient and stable oxygen-reduction reaction over a wide range of pH values. *Chem Eng J* [Internet]. 2021 [cited 2025 Aug 5]. Available from: https://www.sciencedirect.com/science/article/pii/S1385894720330114
263. Alajmi MF, Ahmed J, Hussain A, Ahamad T, Alhokbany N, Amir S, et al. Green synthesis of Fe3O4 nanoparticles using aqueous extracts of Pandanus odoratissimus leaves for efficient bifunctional electro-catalytic activity. *Appl Nanosci (Switzerland)* 2018;8:1427–35.
264. Zhong G, Xu M, Xu S, Fu X, Liao W, Xu Y. Green synthesis of iron and nitrogen co-doped porous carbon via pyrolysing lotus root as a high-performance electrocatalyst for oxygen reduction reaction. *Int J Energy Res*. 2021;45:10393–408.
265. Rosas-Medellín D, Pérez-Salcedo KY, Morales-Acosta D, Rodríguez-Varela FJ, Escobar B. Green synthesis of Pt nanoparticles and their application in the oxygen reduction reaction. *J Mater Res* [Internet]. 2021;36(20):4131–40. Available from: https://doi.org/10.1557/s43578-021-00281-8
266. Zikalala NE, Azizi S, Mpeta LS, Ahmed R, Dube A, Mketo N, et al. Green synthesis of reduced graphene oxide using Persea americana mill. extract: Characterization, oxygen reduction reaction and antibacterial application. *Diam Relat Mater* [Internet]. 2024 [cited 2025 Aug 5]. Available from: https://www.sciencedirect.com/science/article/pii/S0925963524007738
267. Roy P, Ravindranath R, Periasamy AP, Lien CW, Liang CT, Chang HT. Green synthesis of Si-GQD nanocomposites as cost-effective catalysts for oxygen reduction reaction. *RSC Adv* [Internet]. 2016 [cited 2025 Aug 5]. Available from: https://pubs.rsc.org/en/content/articlehtml/2016/ra/c6ra23892k
268. Balasubramanian S, Sheelam A, Ramanujam K, Dhamodharan R. Green, seed-mediated synthesis of Au nanowires and their efficient electrocatalytic activity in oxygen reduction reaction. *ACS Appl Mater Interfaces* [Internet]. 2017 Aug 30 [cited 2025 Aug 6];9(34):28876–86. Available from: https://pubs.acs.org/doi/abs/10.1021/acsami.7b07553
269. Dongre S, Zuccante G, Muhyuddin M, Vecchio CL, Baglio V, Berretti E, et al. Innovative biochar-based electrocatalysts from chilli plants and fruits for sustainable oxygen reduction and hydrogen evolution reactions. *Electrochimica Acta* [Internet]. 2025 [cited 2025 Aug 6]. Available from: https://www.sciencedirect.com/science/article/pii/S0013468625001264

2025 Jun 27 [cited 2025 Aug 6];142:852–63. Available from: https://www.sciencedirect.com/science/article/abs/pii/S0360319925000503?via%3Dihub
300. Stephen AJ, Rees N V., Mikheenko I, Macaskie LE. Platinum and palladium bio-synthesized nanoparticles as sustainable fuel cell catalysts. *Front Energy Res*. 2019;7.
301. Trindade ECA, Antônio R V., Brandes R, de Souza L, Neto G, Vargas VMM, et al. Carbon fiber-embedded bacterial cellulose/polyaniline nanocomposite with tailored for microbial fuel cells electrode. *J Appl Polym Sci* [Internet]. 2020 Sep 15 [cited 2025 Aug 6];137(35):49036. DOI: 10.1002/app.49036. Available from: https://onlinelibrary.wiley.com/doi/10.1002/app.49036
302. Yaqoob AA, Mohamad Ibrahim MN, Yaakop AS. Application of oil palm lignocellulosic derived material as an efficient anode to boost the toxic metal remediation trend and energy generation through microbial fuel cells. *J Clean Prod* [Internet]. 2021 Sep 10 [cited 2025 Aug 6];314:128062. Available from: https://www.sciencedirect.com/science/article/abs/pii/S0959652621022800?via%3Dihub
303. Li H, Gao M, Wang P, Ma H, Liu T, Ni J, et al. Cathode catalyst prepared from bacterial cellulose for ethanol fermentation stillage treatment in microbial fuel cell. *Chinese J Chem Eng* [Internet]. 2021 Dec 1 [cited 2025 Aug 7];40:256–61. Available from: https://www.sciencedirect.com/science/article/abs/pii/S1004954120306510?via%3Dihub
304. Li X, Lv P, Yao Y, Feng Q, Mensah A, Li D, et al. A novel single-enzymatic biofuel cell based on highly flexible conductive bacterial cellulose electrode utilizing pollutants as fuel. *Chem Eng J* [Internet]. 2020 Jan 1 [cited 2025 Aug 7];379:122316. Available from: https://www.sciencedirect.com/science/article/abs/pii/S1385894719317103?via%3Dihub
305. Singh HM, Pathak AK, Chopra K, Tyagi V V., Anand S, Kothari R. Microbial fuel cells: a sustainable solution for bioelectricity generation and wastewater treatment. *Biofuels* [Internet]. 2019 Jan 2 [cited 2025 Aug 7];10(1):11–31. Available from: https://www.tandfonline.com/doi/abs/10.1080/17597269.2017.1413860
306. Magotra VK, Kang TW, Ahmed AT, Inamdar AI, Im H, Ghodake G, et al. Effect of gold nanoparticles laced anode on the bio-electro-catalytic activity and power generation ability of compost based microbial fuel cell as a coin cell sized device. *Biomass Bioenergy* [Internet]. 2021 [cited 2025 Aug 7]. Available from: https://www.sciencedirect.com/science/article/pii/S0961953421002361
307. Sallam ER, Khairy HM, Elnouby MS, Fetouh HA. Sustainable electricity production from seawater using Spirulina platensis microbial fuel cell catalyzed by silver nanoparticles-activated carbon composite prepared by a new modified photolysis method. *Biomass Bioenergy* [Internet]. 2021 [cited 2025 Aug 7]. Available from: https://www.sciencedirect.com/science/article/pii/S0961953421000751
308. Miran F, Mumtaz MW, Mukhtar H, Akram S. Iron oxide-modified carbon electrode and sulfate-reducing bacteria for simultaneous enhanced electricity generation and tannery wastewater treatment. *Frontiers Bioeng Biotechnol* [Internet]. 2021 Nov 12 [cited 2025 Aug 7];9. Available from: https://www.frontiersin.org/articles/10.3389/fbioe.2021.747434/full
309. Duarte KD, Kwon Y. In situ carbon felt anode modification via codeveloping Saccharomyces cerevisiae living-template titanium dioxide nanoclusters in a yeast-based microbial fuel cell. *J Power Sources* [Internet]. 2020 [cited 2025 Aug 7]. Available from: https://www.sciencedirect.com/science/article/pii/S0378775320309551
310. Liu J, Vipulanandan C, Yang M. Biosurfactant production from used vegetable oil in the anode chamber of a microbial electrosynthesizing fuel cell. *Waste Manag* [Internet]. 2017 [cited 2025 Aug 7]. Available from: https://www.sciencedirect.com/science/article/pii/S0956053X17302325
311. Yuan H, Deng L, Qi Y, Kobayashi N, Tang J. Nonactivated and activated biochar derived from bananas as alternative cathode catalyst in microbial fuel cells. *Sci World J*

[Internet]. 2014 Jan 1 [cited 2025 Aug 7];2014(1):832850. DOI: 10.1155/2014/832850. Available from (free full text): https://www.ncbi.nlm.nih.gov/pmc/articles/PMC4163465/

312. Deng L, Yuan Y, Zhang Y, Wang Y, Chen Y, Yuan H, et al. Alfalfa leaf-derived porous heteroatom-doped carbon materials as efficient cathodic catalysts in microbial fuel cells. *ACS Sustain Chem Eng* [Internet]. 2017 Nov 6 [cited 2025 Aug 7];5(11):9766–73. Available from: https://pubs.acs.org/doi/abs/10.1021/acssuschemeng.7b01585
313. Shi YT, Yu YY, Xu ZA, Lian J, Yong YC. Superior carbon belts from Spirogyra for efficient extracellular electron transfer and sustainable microbial energy harvesting. *J Mater Chem A* [Internet]. 2019 [cited 2025 Aug 7]. Available from: https://pubs.rsc.org/en/content/articlehtml/2019/ta/c9ta00332k
314. Huggins TM, Pietron JJ, Wang H, Ren ZJ, Biffinger JC. Graphitic biochar as a cathode electrocatalyst support for microbial fuel cells. *Bioresource Technol* [Internet]. 2015 [cited 2025 Aug 7]. Available from: https://www.sciencedirect.com/science/article/pii/S0960852415007981
315. Ye W, Tang J, Wang Y, Cai X, Liu H, Lin J, et al. Hierarchically structured carbon materials derived from lotus leaves as efficient electrocatalyst for microbial energy harvesting. *Sci Total Environ* [Internet]. 2019 May 20 [cited 2025 Aug 7];666:865–74. Available from: https://www.sciencedirect.com/science/article/abs/pii/S0048969719307983
316. Yuan Y, Yuan T, Wang D, Tang J, Zhou S. Sewage sludge biochar as an efficient catalyst for oxygen reduction reaction in an microbial fuel cell. *Bioresource Technol* [Internet]. 2013 [cited 2025 Aug 7]. Available from: https://www.sciencedirect.com/science/article/pii/S0960852413009942
317. Hung YH, Liu TY, Chen HY. Renewable coffee waste-derived porous carbons as anode materials for high-performance sustainable microbial fuel cells. *ACS Sustain Chem Eng* [Internet]. 2019 Oct 21 [cited 2025 Aug 7];7(20):16991–9. Available from: https://pubs.acs.org/doi/abs/10.1021/acssuschemeng.9b02405
318. Zhu H, Wang H, Li Y, Bao W, Fang Z, Preston C, et al. Lightweight, conductive hollow fibers from nature as sustainable electrode materials for microbial energy harvesting. *Nano Energy* [Internet]. 2014 [cited 2025 Aug 7]. Available from: https://www.sciencedirect.com/science/article/pii/S2211285514001803
319. Zha Z, Zhang Z, Xiang P, Zhu H, Zhou B, Sun Z, et al. One-step preparation of eggplant-derived hierarchical porous graphitic biochar as efficient oxygen reduction catalyst in microbial fuel cells. *RSC Adv* [Internet]. 2021 [cited 2025 Aug 7]. Available from: https://pubs.rsc.org/en/content/articlehtml/2021/ra/d0ra09976g
320. Jiao Y, Hu Y, Han L, Zhou M. Activated carbon derived from rice husk as efficient oxygen reduction catalyst in microbial fuel cell. *Electroanalysis* [Internet]. 2020 Dec 1 [cited 2025 Aug 7];32(12):2969–75. Available from: https://analyticalsciencejournals.onlinelibrary.wiley.com/doi/abs/10.1002/elan.202060409?casa_token=5T6z-COTk5AAAAAA:moka__ORmR8T8oRm6kTia_OYbaRz__uJ2bBQeAEpbeQW4UN8lr-TR21KyhE48RV5q04-gE1jXxCMCjds-
321. Chakraborty I, Sathe SM, Dubey BK, Ghangrekar MM. Waste-derived biochar: Applications and future perspective in microbial fuel cells. *Bioresource Technol* [Internet]. 2020 [cited 2025 Aug 7]. Available from: https://www.sciencedirect.com/science/article/pii/S0960852420308592
322. Zhong K, Li M, Yang Y, Zhang H, Zhang B, Tang J, et al. Nitrogen-doped biochar derived from watermelon rind as oxygen reduction catalyst in air cathode microbial fuel cells. *Appl Energy* [Internet]. 2019 May 15 [cited 2025 Aug 7];242:516–25. Available from: https://www.sciencedirect.com/science/article/abs/pii/S0306261919304593
323. Yuan Y, Liu T, Fu P, Tang J, Zhou S. Conversion of sewage sludge into high-performance bifunctional electrode materials for microbial energy harvesting. *J Mater Chem*

A [Internet]. 2015 [cited 2025 Aug 7]. Available from: https://pubs.rsc.org/en/content/articlehtml/2015/ta/c5ta00458f

324. Wang S, Song W, Hu M, Li F, Song Y, Ju R, et al. Green synthesis of N-doped graphene nanosheets with multi-wrinkled textural properties for boosting oxygen reduction reaction in glucose fuel cell. *Int J Hydrogen Energy*. 2024;72:226–36.
325. Keerthi DS, Vani MM, Tharun S, Teja MRS, Krishnamurthy B. Optimization of pencil graphite electrodes through green-synthesized iron nanomaterials for enhanced biofuel cell performance. *Ionics*. 2025;31:623–39.
326. Elkhrachy I, Singh V, Kumar A, Roy A, Abbas M, Gacem A, et al. Use of biogenic silver nanoparticles on the cathode to improve bioelectricity production in microbial fuel cells. *Front Chem*. 2023;11.

3 Investigation of Corrosion Inhibitors of Natural Compounds Using Tafel Polarization, Electrochemical Impedance Spectroscopy

Melika Sadat Dadgar[1],
Mohammad Hadi Karimi Harandi[2],
Maryam Khayatkashani[3], *and Nasrin Soltani*[4]
[1] Department of Analytical Chemistry, Faculty of Chemistry, University of Kashan, Kashan, Iran.
[2] Department of Chemistry, Imam Hossein University, Tehran, Iran.
[3] Advanced Medical Pharma (AMP-Biotec), Biopharmaceutical Innovation Centre, Via Cortenocera, San Salvatore Telesino, (BN), Italy.
[4] Department of Chemistry, Payam Noor University, Isfahan, Iran.

3.1 INTRODUCTION

Corrosion is a process in which the properties of a material are destroyed by a chemical or electrochemical reaction caused by the environment in which the matter is, and commonly *corrosion* means the electrochemical oxidation of metal. Pure chemical/electrochemical metals and alloys react with a corrosive environment to form a stable compound, and as a resault of this reaction, metal loss occurs. The produced stable compound is called the corrosion product, and the metal surface is corroded. Corrosion is based on the movement of metal ions from the anode electrode to the solution and the passage of electrons from the metal to an electron acceptor. This resaults in an ion current in the solution and an electron current in the metal. The cathodic process requires the presence of an electron receiver, such as oxygen, an oxidizing agent, or hydrogen ions [1, 2].

DOI: 10.1201/9781003229032-3

Considering that the corrosion is a spontaneous process, the annual damage caused by corrosion is a huge cost throughout the world, while in some industries, and even in everyday life, corrosion is used to solve some issues, such as using corrosive chemical solutions to remove layers, deposits, and oxides in industrial units, which is commonly named pickling.

3.2 CORROSION PROTECTION METHODS

There are so many methods to protect metals from corrosion or delay this process, but choosing the most compatible one is not easy. Selecting the corrosion inhibition method depends on the environment and economic conditions. The most common and significant methods are coatings, cathodic protection, anodic protection, and corrosion inhibitors.

3.3 CORROSION INHIBITORS

Corrosion inhibitors are substances that slow down corrosion by affecting cathodic and anodic reactions. These substances are usually effective in low concentrations, and their action is by forming a thin absorbent layer on the metal surface or forming a protective film with or without corrosion products. These are used for the corrosion inhibition of various metals, like soft steel, stainless steel, copper, and aluminum, and are widely used in various industries, storage containers, purifiers, and so on. Acids are used for pickling of ferrous metals and in descaling the surface of metals in the chemical, petroleum, and nuclear industries. Metal corrosion will be followed as a result of being in acid solutions and corrosive environments. Hydrochloric acid is a corrosive solution and widely used in different concentrations as an effective acid for washing pipes and containers and in processing petroleum products. In preventing this issue, corrosion inhibitors are effective. The most important commercial inhibitors used are nitrogen and sulfur compounds, aromatic aldehydes, heterocycles, quaternary nitrogen salts, mercaptans, thioethers, and thiazoles, and among these inhibitors, Schiff bases, with the formula RR'=NR", are known as the best ones. Basically, inhibitors are divided into three main groups: organic, inorganic, and green inhibitors. Here we will discuss organic and inorganic inhibitors briefly, and the main focus will be on green inhibitors.

Inhibitors act by absorbing ions or molecules to the metal surface and reduce the rate of corrosion by one of the following three methods:

1. Reducing cathodic/anodic reaction
2. Reducing the penetration of reactants to the metal surface
3. Increasing the electrical resistance of the metal surface

On the other hand, chemical and natural corrosion inhibitors can be classified as follows based on their inhibition mechanism:

1. Anodic inhibitors
2. Cathodic inhibitors

3. Mixed inhibitors (anodic-cathodic)
4. Surface adsorbents

In general, inorganic inhibitors have cathodic or anodic inhibitory properties. Organic inhibitors have both cathodic and anodic functions and are protected by an absorption film.

3.4 ANODIC INHIBITORS

Anodic inhibitors work through a reducing anodic reaction. That is, through the formation of a film that is absorbed on the metal, they block the anode reaction and cause the metal to passivate.

3.5 CATHODIC INHIBITORS

Cathodic inhibitors prevent the occurrence of metal cathodic reaction due to the presence of alkali metal ions. Therefore, they produce insoluble compounds that are selectively deposited at the cathode sites.

3.6 SURFACE ABSORBENTS

Organic inhibitory compounds sometimes act as cathodic, anodic, or mixed anodic-cathodic inhibitors; however, they mostly inhibit corrosion through a surface adsorption process that forms as a film. The inhibition mechanism of organic compounds in acidic environments is mostly based on creating a bond between these compounds and the metal surface. Because in acidic environments, the metal surface is usually oxide-free, these bonds are created through electrostatic interactions between charged metal surfaces and charged organic compounds and, as a result, cause surface adsorption. Another mechanism is based on electron transfer between inhibitor molecules (which have aromatic rings, double or triple covalent bonds, and lone pair electrons) and the metal substrate.

Organic compounds can be divided into two general categories based on their performance in corrosion inhibition, which are hydrophilic and nonpolar hydrophobic functional groups. *Hydrophobic* structures prevent corrosive species from entering the metal surface, while *hydrophilic* structures (such as mercapto, amine, hydroxyl, phosphate, carboxylic, etc.) cause bonding with the metal surface.

3.7 ORGANIC CORROSION INHIBITORS

Organic inhibitors are often used in acidic environments. Their effectiveness depends on the chemical composition, molecular structure, and interaction of the molecule with the metal surface.

Organic compounds containing electronegative functional groups, π electrons present in conjugated double bonds and triple bonds, generally show good inhibitory properties by providing electrons through π electrons. The interaction between the functional groups and the surface of the metal through the nitrogen-, sulfur-, and oxygen-free pairs of electron plays an important role in protecting metals from corrosion. Both mentioned factors increase the inhibitory properties of inhibitory compounds in organic corrosion inhibitors.

3.8 INORGANIC CORROSION INHIBITORS

Inorganic corrosion inhibitors are used in neutral conditions and affect anodic reactions. Chromates, nitrates, and molybdates are parts of this group which form film and change potential toward positive values in pH 6–9 to inhibit corrosion.

3.9 GREEN CORROSION INHIBITORS

As explained, *inhibitors* are substances that are widely used in various industries, especially in acidic environments, to prevent the corrosion of metals. Usually, the most important commercial inhibitors used are aromatic aldehydes, heterocycles, quaternary nitrogen salts, mercaptans, thioethers, and thiazoles, which have aromatic rings and nitrogen and sulfur groups in their structure. The very good performance of these materials in the protection of metals has attracted the attention of the industry and has led to the widespread use of these materials. But the biggest problem of these materials is that most of these are toxic and harmful to the environment and to living organisms. In recent years, with increasing awareness of the environment and of green chemistry, many works have been done to use environmentally friendly materials instead of harmful artificial chemicals. For this reason, researchers have made a lot of effort in the field of identifying and investigating natural substances with similar structures to effective inhibitors. It should be mentioned that the first corrosion inhibitors that were used in the beginning were naturally substances that were extracted from different parts of plants. In the following, various parts of the structure of natural compounds as green corrosion inhibitors and their mechanism of action will be investigated.

3.10 FRUIT

Dehghani et al. evaluated the effect of aqueous *Citrullus lanatus* fruit (CLF) extract as a green corrosion inhibitor on mild steel surface in 1 M hydrochloric acid solution. The extract used contained several plant chemicals (for example, citrulline, hesperetin, resveratrol, and riboflavin) that played an important role in the metal corrosion inhibition method, and the presence of these main plant chemicals was determined by FT-IR spectroscopy along with UV-visible analysis. Corrosion inhibition properties were investigated using electrochemical (EIS and polarization) and surface morphological (SEM, AFM, contact angle) techniques. EIS analysis showed a significant corrosion inhibition degree (91%) using 800 ppm CLF extract. According to the results of the polarization test, the corrosion inhibition of CLF extract has occurred by the postponing of the cathodic reaction. Also, through SEM/AFM, graphical results showed that the interactions of CLF extract molecules with the steel surface have a great effect on preventing surface damage. Finally, the data obtained from molecular simulation and quantum mechanical analysis showed the absorption of CLF extract compounds on the steel surface [3].

Cherrad et al. evaluated the effect of *Cupressus arizonica* fruit essential oil (CAFEO) as a green corrosion inhibitor for carbon steel (CS) in 1 M HCl solution using electrochemical techniques, for example, potentiodynamic polarization and electrochemical impedance spectroscopy. Also, SEM was used to examine the morphology of the metal surface. The used extract contained three main plant chemicals, including a-pinene (51.07%), myrcene (17.92%), and limonene (9.66%). Essential oils with a higher percentage of a-pinene have significant corrosion inhibitory properties. Experimental results showed that CAFEO, as a corrosion inhibitor, protects carbon steel well in a 1 M HCl solution. Potentiodynamic polarization showed that the inhibition of CAFEO corrosion is by postponing the cathodic reactions. The EIS data showed that the charge transfer resistance of the CS electrode in the blank solution increased compared to the HCl solution containing 0.5 g/L CAFEO inhibitor at 298 K. In this condition, the inhibition efficiency was equal to 93%. The results showed that increasing temperature has a negative effect on the corrosion inhibition efficiency of CAFEO inhibitor, so that at 323 K, the inhibition efficiency of CAFEO decreased to 77%. Also, SEM graphic results showed that the CAFEO inhibitor inhibits the corrosion of carbon steel by creating a protective barrier against acid attack. Density functional theory (DFT) was used to perform theoretical calculations of the reactivity of three compounds: a-pinene, myrcene, and limonene [4].

Aqueous and alcoholic extracts of *Garcinia cambogia* fruit peel (GCW and GCA) were used to investigate the corrosion of mild steel in 0.5 N HCl medium by Rani et al. Polarization and electrochemical impedance spectroscopy methods were used in this study. *Garcinia cambogia* fruit is environmentally friendly and affordable, so it is considered as a green corrosion inhibitor. GCW and GCA contain six main plant chemicals, including cyanidin, garcinol, hydroxycitric acid, malic acid, ascorbic acid, and tartaric acid, and due to the presence of these organic acids, its pH is slightly acidic and around 5. IR and UV spectral observations confirmed the presence of these different compounds in the structure of GCW and GCA. These results confirmed that these compounds – GCW, rich in cyanidin, and GCA, mainly rich in hydroxy citric acid, along with a smaller portion of cyanidin – cause corrosion inhibition properties in these extracts. EIS analysis showed inhibition efficiency of 91.50% and 92.85% for 4 (v/v%) GCW and GCA, respectively, at room temperature and confirmed the better inhibition ability of GCA than GCW in acidic environment. SEM studies also confirmed these results and showed that the adsorption of the active species of the extracts on the metal surface is more favorable for GCA. According to the results of the polarization test, the corrosion inhibition of GCW is by postponing the cathodic hydrogen evolution reaction, while for GCA, the corrosion inhibition property is by resisting the metal anode dissolution reaction. Based on the Langmuir adsorption isotherm, the molecules of both *Garcinia cambogia* fruit peel and extracts reacted with the mild steel surface, and as a result, the steel was covered with a layer. Also, theoretical studies showed that the active compounds GCW and GCA are adsorbed on the metal surface through coordinate-type interaction and electron back-donation from the metal to the inhibitor molecules [5].

Thomas et al. investigated the effect of *Garcinia indica* fruit peel extract (GIW) as a green corrosion inhibitor for mild steel in a 1 M HCl solution at four different temperatures using the synergistic interaction of the inhibitor and potassium iodide (KI). Weight loss method and electrochemical methods, including polarization and electrochemical impedance spectrum, were used for this investigation. GIW contains the main natural compounds of cyanidin anthocyanins, including cyanidin 3-glucoside, cyanidin sambubioside, and hydroxycitric acid, which causes corrosion inhibition in this extract. EIS analysis showed that by adding 200 ppm KI to 4% GIW extract, inhibition efficiency increases from 89% to 90% at 303 K, but this increase will be from 74% to 81% at 333 K. The absorption process is mainly the physical absorption of the GIW/KI pair. The increase in binding energy upon the adsorption of GIW and KI was confirmed by Monte Carlo simulations of the constituent molecules [6].

Liao et al. investigated the effect of extract from the waste of lychee fruit as a green corrosion inhibitor for mild steel in a 0.5 M HCl solution. In this study, weight loss method and electrochemical methods including polarization and electrochemical impedance spectrum were used. The natural compounds in lychee fruit extract, (-)-epicatechin, (-)-epigallocatechin, and cyanide-3-glucoside, have the capacity of electron donation (donating π electron and lone pair electron) and strong chelation with iron, which cause corrosion inhibition in this extract. EIS analysis showed a significant corrosion inhibition degree (91%) using 800 ppm CLF extract. Electrochemical measurements and weight loss showed that the best inhibitory effect of the ethanol extract of lychee's peel and seed (EELPS) is obtained using 600 mg/L EELPS in 1.5 hours. In this case, the degree of corrosion inhibition is 97.95%. Adsorption sites and absorption behavior of natural compounds were investigated using theoretical calculations. The results confirmed the presence of hydroxyl groups and phenol rings in these natural compounds and stated that EELPS has good inhibitory properties [7].

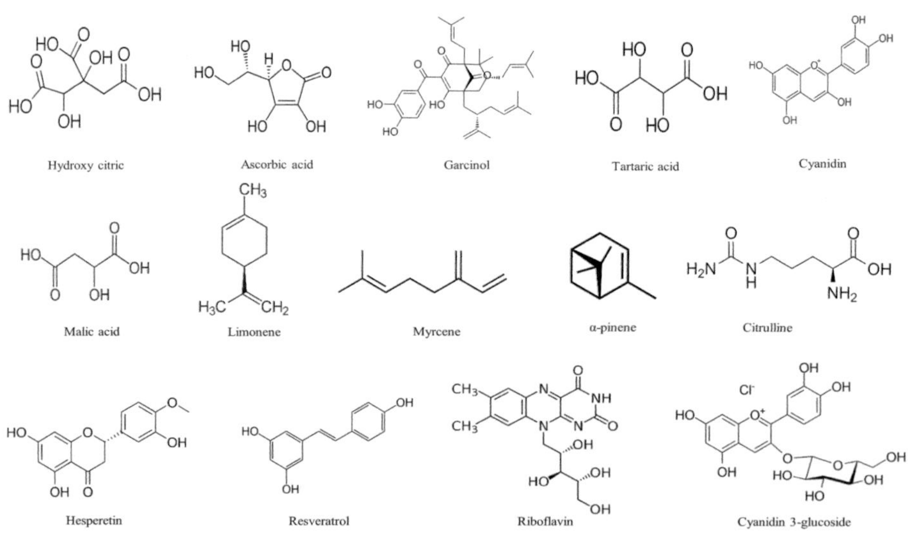

FIGURE 3.1 The active ingredients in fruits of plants applied as corrosion inhibitors.

3.11 LEAF

Anupama et al. investigated the corrosion inhibitory properties of extracts of the leaves of *Pimenta dioica* (PD) as mild steel inhibitors in 0.5 M and 1 M HCl. PD extracts contain the main natural compounds eugenol, methyl eugenol, β-caryophyllene, myrcene, limonene, and 1,8-cineole, which cause corrosion inhibition in them. Corrosion rate was evaluated at three different temperatures using electrochemical impedance spectroscopy and polarization techniques. Potentiodynamic polarization showed that PD extracts act as mixed-type inhibitors at all these temperatures. EIS data showed that the best inhibition efficiency of PD extracts is at 303 K and is 95%. Adsorption studies were also conducted to investigate the mechanism of corrosion inhibition. Based on the Langmuir adsorption isotherm, the molecules of PD extracts reacted with the mild steel surface, and as a result, the steel was covered with a layer [8].

Olasehinde et al. synthesized titanium nanocomposite from the methanolic extract of *Chromolaena odorata* leaves and investigated the effect of the titanium–*Chromolaena odorata* $\left(Ti-CO\right)$ nanocomposite as a mild steel corrosion inhibitor in 0.5 M and 1 M HCl acid environments. The weight loss method and electrochemical methods, including polarization and electrochemical impedance spectrum, were used to investigate the corrosion inhibitor performance of Ti-CO. Weight loss measurements showed that for 1.0 g/L of Ti-CO inhibitor, the inhibition efficiency was 92.39%, while the inhibition efficiency was 84.77% for the methanolic extract of *Chromolaena odorata* leaves under the same conditions. The inhibition efficiency ofTi-CO increased with increasing inhibitor concentration but decreased with increasing temperature. EIS data showed that the charge transfer resistance of the electrode increased in the presence of the Ti-CO inhibitor, which is due to the adsorption of Ti-CO on the metal surface. Based on the Langmuir adsorption isotherm, Ti-CO inhibitor molecules reacted with the surface of mild steel, and based on this, the physical adsorption of the inhibitor on the surface of mild steel was predicted. SEM graphical results also confirmed the formation of a protective layer ofTi-CO on the mild steel surface [9].

Nathiya et al. used two different solvent extracts of *Dryopteris cochleata* leaves (DCLME and DCLWE) as aluminum corrosion inhibitors in 1 M H_2SO_4. The weight loss method and electrochemical methods were used to check the corrosion inhibitor performance of DCLME and DCLWE extracts. The methanolic extract of *Dryopteris cochleata* leaves was a better inhibitor than the aqueous extract. Based on the graphic results, the mechanism of action of DCLME and DCLWE extracts is in the form of molecular adsorption on the surface of aluminum. Potentiodynamic polarization showed that these two extracts act as mixed-type inhibitors at all temperatures. EIS data showed that the best inhibition efficiency is in the presence of 2,400 ppm of DCLME and DCLWE extracts and is 95.09% [10].

Dehghani et al. investigated the corrosion inhibitory properties of eucalyptus leaves extract (ELE) as mild steel inhibitor in 1 M HCl. Electrochemical methods including polarization and electrochemical impedance spectroscopy were used to check the corrosion inhibitor performance of ELE. EIS data showed that the best inhibition efficiency is in the presence of 800 ppm ELE and is equal to ~88%. It was also shown that increasing the concentration of ELE inhibitor leads to a significant increase in charge transfer resistance. Potentiodynamic polarization showed that the

ELE extract acts as a mild cathodic inhibitor. Based on the Langmuir adsorption isotherm, ELE inhibitor molecules reacted with the surface of mild steel, and based on this, the physical adsorption of the inhibitor on the surface of mild steel was predicted. The results of the molecular simulation confirmed this theory [11].

Qiang et al. investigated the corrosion inhibitory properties of Ginkgo leaf extract (GLE) as an environmentally friendly corrosion inhibitor for X70 steel in 1 M HCl. GLE extract contains the main natural compounds isorhamnetin, sciadopitysin, 6-hydroxykynurenic acid, and 4-O-methylpyridoxine, which cause its anticorrosion properties. Electrochemical methods were used to investigate the corrosion inhibitor performance of GLE. Based on the EIS analysis, inhibition efficiency in the presence of 200 mg/L of GLE is higher than 90% at all temperatures, so that the best inhibition efficiency of GLE was obtained at 318 K and equal to 92.5%. SEM graphical results also showed that the excellent inhibitory property of GLE is related to the formation of inhibitory-absorbing layers on the surface of X70 steel. Based on the Langmuir adsorption isotherm, GLE molecules react with the X70 steel surface, and as a result, the steel is covered with a layer [12].

Mobin et al. investigated the corrosion inhibitory properties of *Lagerstroemia speciosa* leaf extract (LSLE) as an environmentally friendly corrosion inhibitor for mild steel in 1 MHCl by weight loss method and electrochemical methods, including polarization and electrochemical impedance spectroscopy. Based on the EIS analysis, LSLE active groups cause the formation of an inhibitory film on the metal surface. Weight loss measurements showed that for 500 ppm of LSLE inhibitor at 333 K, the inhibition efficiency is higher than 94%. The results of potentiodynamic polarization showed that the LSLE inhibitor is a type of mixed inhibitor with anodic predominance. Based on the Langmuir adsorption isotherm, LSLE molecules react with the mild steel surface, and the physical adsorption of the inhibitor on the mild steel surface was predicted. Also, the UV-vis results showed that the Fe-LSLE complex is formed during the corrosion reaction [13].

Muthukrishnan et al. investigated the corrosion inhibitory properties of *Lannea coromandelica* leaf extract (LCLE) as a green corrosion inhibitor for mild steel in 1 MH_2SO_4 by weight loss method and electrochemical methods, including polarization and electrochemical impedance spectroscopy. The results showed that the inhibition efficiency of LCLE increases with increasing concentration but decreases with increasing temperature. The results of potentiodynamic polarization showed that the LCLE inhibitor is a type of mixed inhibitor. Based on the EIS analysis, with the increase of LCLE concentration, the charge transfer resistance increases and the capacity of the double layer capacitor decreases, and on this basis, inhibition efficiency of 93.8% was obtained. Based on the Langmuir adsorption isotherm, LCLE molecules react with the mild steel surface, and physical adsorption of the inhibitor on the mild steel surface is predicted. FT-IR, XRD, SEM and AFM techniques also confirmed this physical absorption [14].

Rahal et al. investigated the corrosion inhibitory properties of olive leaf extract as a green corrosion inhibitor for pure copper in 0.5 MNaCl by weight loss and electrochemical methods. Based on EIS analysis, at the highest inhibitor concentration (2.42 mmol L^{-1}), the inhibition efficiency reached 90% after 24 hours of immersion. Polarization curves about the inhibition process confirmed that olive leaf extract is a type of cathodic inhibitor and prevents the transport, and thus the reduction, of dissolved O_2. The results of HPLC showed that oleuropein is the main compound in olive leaf extract, so this compound probably causes corrosion inhibitory properties in the examined extract [15].

Zhou et al. investigated the corrosion inhibitory ability of *Syzygium samarangense* leaf extract (SSLE) as an environmentally friendly corrosion inhibitor for Q235 steel in 0.5 M H_2SO_4 by electrochemical methods. Graphical results of surface morphology show that Q235 steel becomes smooth and shiny in the presence of an SSLE inhibitor. The results showed that for 400 mg/L SSLE inhibitor at 298 K, the inhibition efficiency is 90%. Potentiodynamic polarization showed that SSLE acts as a mixed-type inhibitor and can effectively limit cathodic and anodic reactions. Computational data also confirmed the corrosion inhibition performance of SSLE. Based on the Langmuir adsorption isotherm, SSLE molecules are adsorbed at the interface of Q235 steel and the solution, thereby inhibiting corrosion [16].

Fernandes et al. used the ability of tropical almond (*Terminalia catappa*) leaf extract to inhibit corrosion as an antioxidant additive to soybean oil biodiesel to investigate the corrosion improvement of zinc and carbon steel 1020 by gravimetric method. The weight loss measurement showed that this green corrosion inhibitor can effectively inhibit the corrosion of carbon steel 1020, and in this case, inhibition efficiency of 78.41% was obtained. However, in the case of mild zinc, some corrosion was observed at room temperature (78.41% inhibition efficiency), which was intensified by increasing the temperature to 60°C. The results showed that, at room temperature, a greenish protective layer is created on the surface of both metals from the antioxidant components in the plant, which prevents corrosion in both metals [17].

In a study, da Silva et al. investigated the corrosion inhibitory effect of *Syzygium cumini* leaf extract (SCLE) for carbon steel in 0.5 MH_2SO_4 by electrochemical methods. The results showed that inhibition efficiency increases with the increase in SCLE concentration, and this increase in efficiency is related to the reduction of cathodic currents and the blocking of corrosion sites on the carbon steel surface. Adsorption studies were also conducted to investigate the mechanism of corrosion inhibition. Based on the Langmuir adsorption isotherm, SCLE molecules reacted with the carbon steel surface, and as a result, the steel was covered with a layer. The results of SEM and contact angle measurements also confirmed the inhibiting performance of SCLE in reducing carbon steel corrosion in an acidic environment [18].

During a study, Harb et al. studied the corrosion inhibitory effect of Olive leaf extract for mild steel in 0.1 M NaOH + 0.5 M NaCl with electrochemical methods and Mott–Schottky analyses. Different solvents, including methanol, ethyl acetate, dichloromethane, and hexane, were used for extraction, and the results showed that the methanolic extract has corrosion inhibitory properties of 91.9%. The inhibitory efficiency of the extract increases by increasing the polarity of the extraction solvents. Potentiodynamic polarization showed that the olive leaf extract acts as a mixed-type inhibitor with a dominant anodic effect. This extract contains various phenolic and flavonoid compounds, including diisooctyl phthalate, hexanedioic acid, mono(2-ethylhexyl) ester, 9-octadecenamide, (Z)-, oxiraneoctanoic acid, 3-octyl-, methyl ester, cis, oleic acid, 1-eicosene, 1, 2-benzenedicarboxylic acid, butyl 8-methylnonyl ester, n-hexadecanoic acid, benzene, (1-methyldodecyl)-, phenethylamine, 3-benzyloxy-2-fluoro-b-hydroxy-, 1-Octadecanol, Phenol,2,6- bis(1,1-dimethylethyl)-4-ethyl-,

Tetradecanoicacid, 1,7-di-iso-propylnaphthalene, cyclopentaneacetic acid, 3-oxo-2-pentyl-, methyl ester, and benzophenone. These compounds contain N and O heteroatoms with π electrons and cause corrosion inhibition in this extract [19].

In another study, Soltania and Khayatkashani investigated the corrosion inhibition effect of *Malva sylvestris* leaf extract for mild steel in 2.0 MHCl by weight loss method and electrochemical methods. The effect of temperature on the behavior of the extract was investigated by using the potentiodynamic polarization method. The results showed that the inhibition efficiency increases with increasing concentration of the extract, so that the highest inhibition efficiency corresponds to the concentration of 2.0 g/L of the extract and is equal to 93%. Based on the Langmuir adsorption isotherm, *Malva sylvestris* leaf extract molecules reacted with a mild steel surface, and physical adsorption was observed as a result [20].

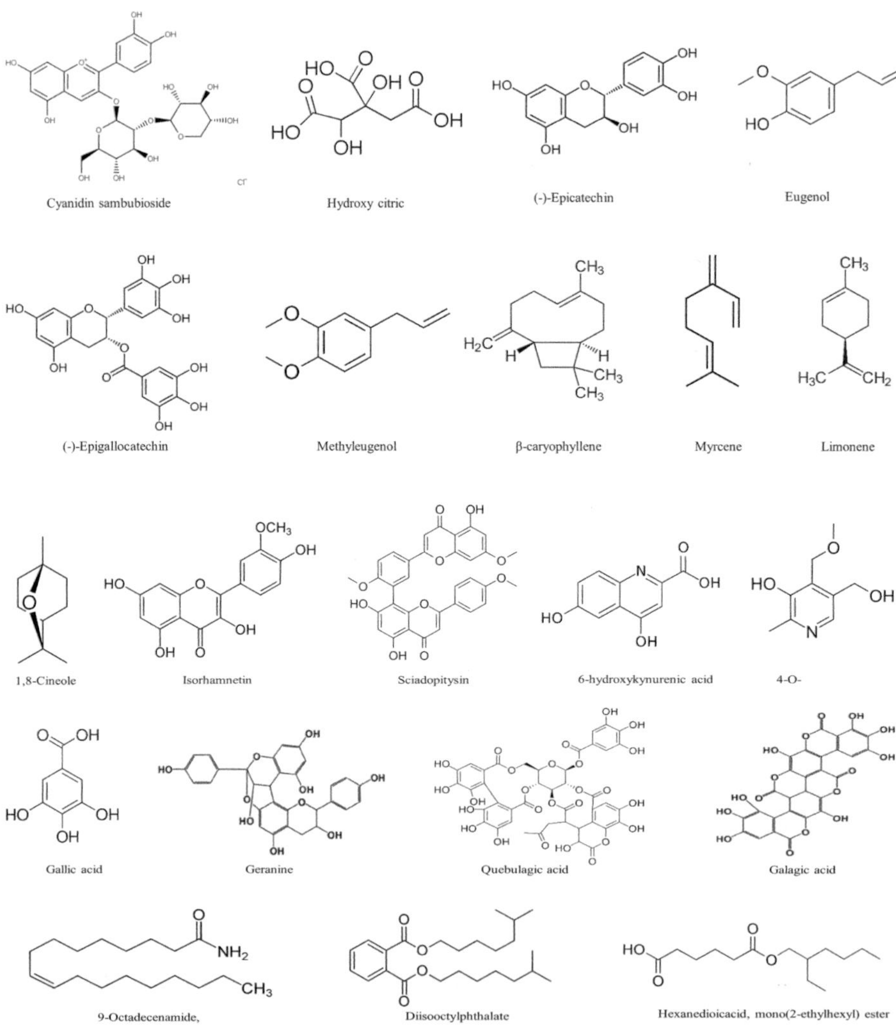

FIGURE 3.2 The active ingredients in the leaf of plants applied as corrosion inhibitors.

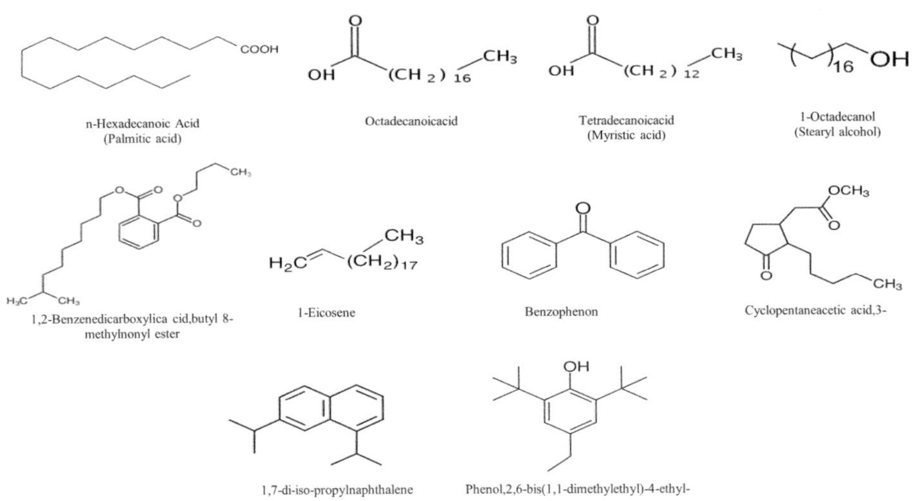

FIGURE 3.2 (Continued)

3.12 SEED

Hamdani et al., investigating on the corrosion inhibitory effect of alkaloid extracts of *Retama monosperma (L.) Boiss.* seeds (AERS), studied its effect on carbon steel in 1 M HCl with electrochemical and surface characterization techniques. Using the potentiodynamic polarization method, the effect of temperature on the behavior of the extract was investigated. The results showed that AERS shows good inhibitory properties, so that the inhibition efficiency for 400 mg/L AERS at 30°C temperature was 94.4%. Based on Langmuir adsorption isotherm, AERS molecules reacted with a carbon steel surface, and as a result, physical adsorption was observed. Potentiodynamic polarization showed that AERS acts as a mixed-type inhibitor. EIS data showed that adding AERS to an HCl solution decreases the charge capacity and increases the charge/discharge performance of the interface and helps form an adsorbed layer on the steel surface. The data obtained from XPS also confirmed that the inhibitory property of AERS molecules is related to their physical absorption on the metal surface, and this inhibitory layer consists of a mixture of iron oxide/hydroxide [21].

In another study, Galliaa et al. investigated the corrosion inhibition effect of Pehuén (*Araucaria araucana*) seed extract for steel in 1.0 M HCl by weight loss method. Using the potentiodynamic polarization method, the effect of temperature on the behavior of the extract was investigated. The results showed that the extract of Pehuén shows good inhibitory properties, so that the inhibition efficiency for 400 μg dried coat of Pehuén per mL of HCl was 61%. Pehuén extract contained the main natural compounds gallic acid, catechin, quercetin, and tannin, which inhibit corrosion in them [22].

In another study, Damej et al. investigated the corrosion inhibitory effect of *Cannabis sativa L.* seed oil (CSL) for E24 carbon steel in 1 M HCl by weight loss method and electrochemical methods. Gas chromatography analysis showed that CSL contains natural compounds, including linoleic acid (51.3%), oleic acid (20.3%), α-linolenic acid (15.7%), and palmitic acid (7.9%). The gravimetric and electrochemical results showed that the inhibition efficiency for 1 g/L CSL is 92.4%. Potentiodynamic polarization showed that the CSL extract acts as a mixed-type inhibitor with cathodic predominance at all these temperatures. SEM graphic results showed the formation of a protective layer on the surface of E24 carbon steel, which prevents corrosion [23].

In another investigation, Abbout et al. investigated the corrosion inhibition and absorption behavior of *Ceratonia siliqua L* seeds (CFE) extract for carbon steel in 1 M HCl using different methods, including electrochemical methods and measurement of thermodynamic and kinetic parameters. Gas chromatography analysis showed that the CFE extract contains natural compounds, including vanillin (27.78 wt.%), α-tumerone (25.69 wt.%), 2-heptanal (11.68 wt.%), mome-inositol (9.49 wt.%), 1-dotiacontanol (4.94 wt.%), and linoleic acid (2.92 wt.%). The results show that the CFE extract has a suitable anticorrosion effect, so that the inhibition efficiency for 100 mg/L CFE at a temperature of 298 Kelvin is 95%. This inhibitory property is probably due to the absorption of CFE extract on the surface of carbon steel, which was also confirmed by SEM graphic results. According to the Langmuir adsorption isotherm, CFE extract molecules react with carbon steel surface, and no interaction between CFE molecules takes place during adsorption. Computational chemistry calculations were also performed to better understand the inhibition and adsorption behavior of the main components of CFE extract on the Fe (110) surface [24].

In another study, Dehghani et al. investigated the corrosion inhibition and absorption behavior of aqueous clove seed extract for mild steel in 1 M HCl using electrochemical, along with surface morphology, and theoretical molecular simulation related to density functional theory, techniques. Potentiodynamic polarization showed that clove seed extract acts as a mixed-type inhibitor, and therefore, the corrosion inhibitory effect occurs in it through the delay in cathodic and anodic combined reactions with low cathodic prevalence. EIS data showed that the inhibition efficiency in the presence of 800 ppm of clove seed extract is 93%, which is probably related to the absorption of the inhibitor on the surface of mild steel. SEM graphic images also confirm these results. Adsorption studies were also conducted to investigate the mechanism of corrosion inhibition. Based on the Langmuir adsorption isotherm, clove seed extract molecules reacted with the surface of mild steel, and as a result, the steel was covered with a layer [25].

In another research study, Tabatabaei et al. studied the corrosion inhibition and absorption behavior of aqueous extract of Esfand seed (ESE) for carbon steel in 1 M HCl using weight loss, EIS, and polarization techniques. In addition, the adsorption of ESE on the carbon steel surface was investigated by Monte Carlo (MC), molecular dynamics (MD), and quantum mechanics (QM) methods. EIS data showed that in addition to the solution containing ESE, zinc cations also increase the corrosion efficiency. The highest inhibition efficiency is related to a solution containing 300 ppm ESE + 700 ppm zinc, so that the inhibition efficiency of this sample reached 98.8% after 264 hours. MC and MD simulations also support the adsorption of ESE as an inhibitor on the carbon steel surface [26].

In another study, Haldhar et al. investigated the corrosion inhibition and absorption behavior of *Gloriosa superba* seed extract for carbon steel in 1 M H_2SO_4 using weight loss, EIS, and polarization techniques. Potentiodynamic polarization showed that *G. superba* extract acts as a mixed-type inhibitor. Based on the results, in the presence of 700 mg/L of *G. superba* extract, the inhibition efficiency is equal to 93.84%. SEM and AFM were used to detect the formation of a protective layer of *G. superba* extract on carbon steel and protect it against corrosion. Based on the Langmuir adsorption isotherm, the molecules of *G. superba* extract reacted with the carbon steel surface, and as a result, the steel was covered with a layer. Also, FT-IR spectra showed the presence of beneficial aggregates and different heteroatoms in the inhibitor. All the results confirm that *G. superba* creates a protective layer against corrosion [27].

In another study, Njoku et al. investigated the corrosion inhibition and adsorption behavior of *Piper guineense* seed extract (PGS) for Q235 carbon steel in acid chloride medium (0.25 M H_2SO_4 + 0.5 M NaCl) using weight loss and EIS techniques. Potentiodynamic polarization showed that the PGS extract acts as a mixed inhibitor, so it has corrosion inhibition effects simultaneously in both cathodic and anodic reactions. EIS data showed that phytochemical compounds of PGS have strong surface adsorption on the Q235 carbon steel surface. Based on the results, in the presence of 0.5 g/L of PGS extract, inhibition efficiency >98% is achieved for long periods of time (up to 144 hours). Based on the Langmuir adsorption isotherm, PGS extract molecules react with the carbon steel surface, and as a result, the steel is covered with a surface coating layer containing the extract components. The large negative values of the calculated absorption energies confirm the significant anticorrosion performance of the PGS extract [28].

In another study, Wang et al. investigated the corrosion inhibition and adsorption behavior of grape seed proanthocyanidins (GSP) for Q235 steel in seawater using weight loss technique. Structural instability of GSP is their main problem. Wang et al. solved this problem by using zein–sodium caseinate (NaCas) composite nanoparticles loaded with GSP. The stability test showed that the particles have good corrosion inhibition efficiency stability within 15 days in a wide range of ionic strength and pH value, and therefore, they can be suitable and compatible for use in seawater environment. The results showed that zein-NaCas composite nanoparticles loaded with 200 mg/L GSP for Q235 steel in seawater have higher inhibition efficiency than GSP (96.06% vs. 71.89%) [29].

In another research work, Mostafatabar et al. investigated the corrosion inhibition and absorption behavior of *Heracleum persicum* seed extracts (HPSE) for mild steel in 1 M HCl using experimental (electrochemical and surficial studies) and theoretical (computer simulation) methods. Potentiodynamic polarization showed that the HPSE extract acts as a mixed inhibitor (anodic/cathodic) in acidic solution, so it has corrosion inhibitory effects simultaneously in both cathodic and anodic reactions. EIS data showed that HPSE compounds have strong surface adsorption on mild steel surface. Based on the weight loss results, in the presence of 400 ppm HPSE, 96% inhibition efficiency was obtained after 5 hours of immersion at 25°C in acidic solution. Based on the FE-SEM and AFM results, a smoother surface was obtained, which indicates the production of HPSE protective layers on the mild steel substrate. Also, the CA test confirmed the increase in the hydrophobicity of the metal surface in the presence of HPSE, indicating the adsorption of organic molecules such as pimpinellin and bergapten at the mild steel/solution interface. The Langmuir isotherm also confirmed the experimental data, indicating a suitable protective performance of HPSE. The suggested theoretical findings support the surface adsorption of HPSE inhibiting chemicals through reactive centers [30].

Also, metals can be protected in the acidic corrosive environment of HCl by using the seed extract of lemon [31], mustard [32], *Ocimum basilicum* [33], *Peganum harmala* [34], *Phyllanthus emblica* [35], quince [36], quinoa [37], and *Retama monosperma (L.) Boiss* [38].

To prevent the corrosion of aluminum alloys, including 2024-T3 and 7075-T6, in alkaline NaCl environment, the extract of *Ocimum basilicium* seeds (OBS) [39] and pumpkin seeds (PS) [40] can be used, respectively.

In another study, Rehioui et al. used 1,000 ppm of *Opuntia dillenii* seed oil containing linoleic acid to prevent the corrosion of iron in acid rain and obtained an inhibition percentage of 99.19% [41].

In another research work, Souzaa et al. obtained more than 99% inhibition efficiency on P110 carbon steel in CO_2-saturated brine using passion fruit seed oil [42].

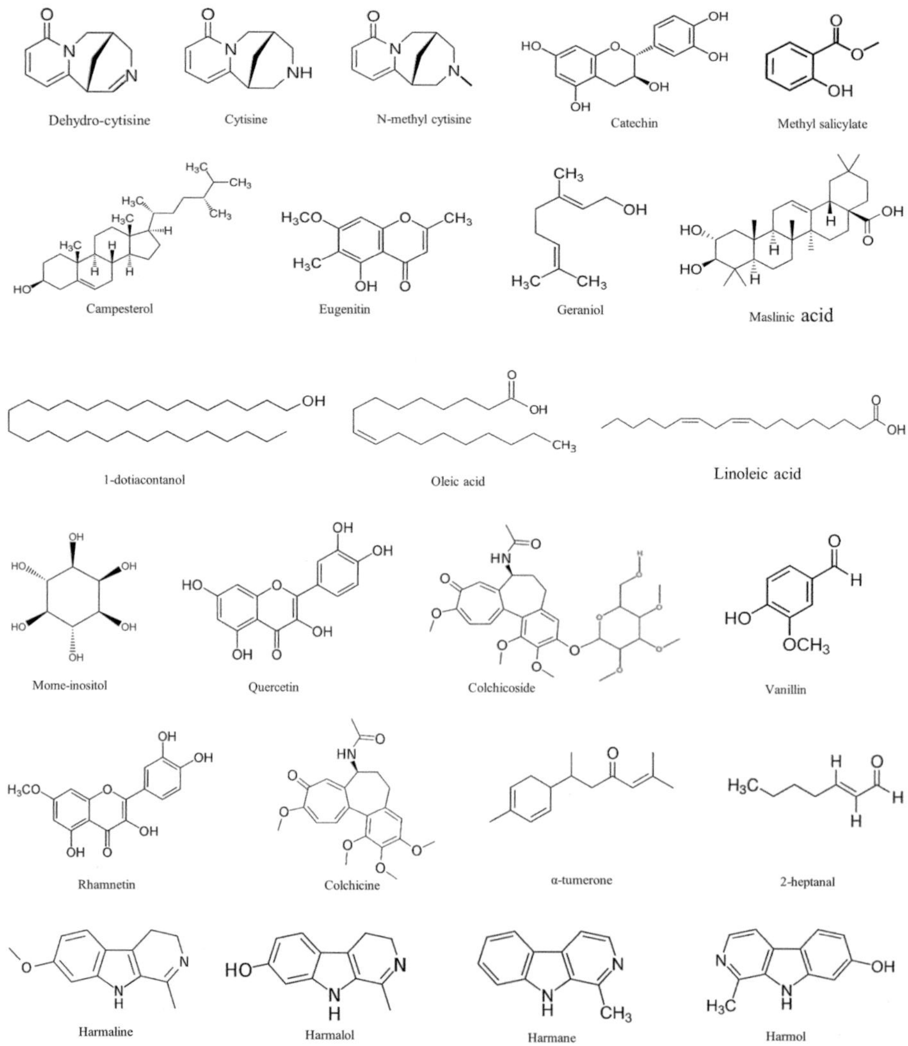

FIGURE 3.3 The active ingredients in the seed of plants applied as corrosion inhibitors.

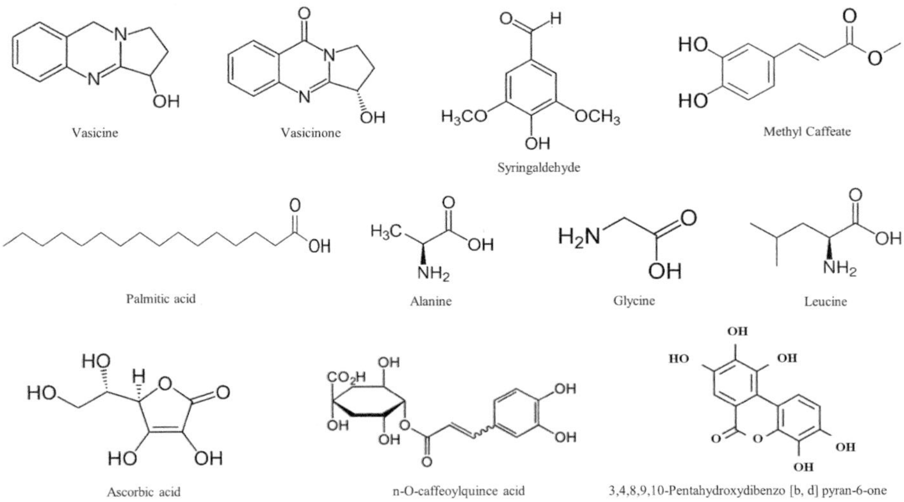

FIGURE 3.3 (Continued)

3.13 WASTE

In another research work, Fekri et al. investigated the corrosion inhibition and absorption behavior of turnip peel extract (TPE) for copper in 3.5% NaCl using weight loss, EIS, and polarization methods. Based on GC-MS analysis, the TPE inhibitor contains many natural compounds, including dimethyl trisulfide, 5-cyano-1-pentene, diethoxy methyl acetate, 3,5-dihydroxy-6-methyl-2,3-dihydropyran-4-one, desulphosinigrin, and benzene is propane nitrile, which creates corrosion inhibition properties in it. It was found that by increasing the amount of TPE, the inhibition efficiency increases, so that in the presence of 20% V/V of TPE, the maximum IE equal to 92% was obtained. It was also found that the inhibition efficiency of TPE decreases with increasing temperature. Based on the Langmuir adsorption isotherm, TPE molecules react with the copper surface, and physical adsorption is observed as a result [43].

Also, metals can be protected in the acidic corrosive environment of HCl by using waste extract, including sugar beet pulp pectin [44], the hard shell of the pistachio nut [45], *Akebia trifoliata Koidz.* peel [46], almond [47], carrot pomace [48], citrus limetta pulp [49], tea factory [50], and waste rice husk [51].

In another research work, Sanni et al. investigated the corrosion inhibition and adsorption behavior of palm kernel shell–derived extracts (PKS) for stainless steel in 0.5 M H_2SO_4 using weight loss and polarization methods. SEM graphical data confirmed the protective function of the PKS extract as an inhibitor. The obtained results show that the PKS extract can suppress acid corrosion in all concentration ranges. Potentiodynamic polarization showed that the PKS extract acts as a mixed-type inhibitor, and in the presence of 4 g of this extract, 93% inhibition efficiency is obtained. Based on the Langmuir adsorption isotherm, PKS extract molecules react with the stainless-steel surface, and as a result, the steel is covered with a surface coating layer containing the extract components. Therefore, PKS extract on the metal surface increases the mass and charge barrier and thus reduces the corrosion rate [52].

In another study, Vorobyova et al. investigated the inhibitory behavior of apricot pomace extract (APE) for mild steel in 0.5 M NaCl using weight loss method. GC-MS results showed that the extract contains a large number of natural compounds, including benzaldehyde, gallic acid, caffeic acid, chlorogenic acid, thymol, linalool, β-cyclocitral, catechin, and terpineols. Weight loss results showed that in the presence of 0.5 mg/L of this extract, the inhibition efficiency is 89.06%. The morphological study confirmed the adsorption of APE molecules on the mild steel surface [53].

In another study, Gao et al. investigated the inhibition behavior of tannic acid–grafted cornstalks–derived carbon nanosheets (TA/CNS) and 8-hydroxyquinoline-loaded mesoporous silica (8-HQ/MS) for steel in 3.5 wt% NaCl using the method electrochemical studies were carried out. EIS results showed that TA/CNS has good anticorrosion performance, and the coating has self-healing properties. The high performance and long life of the coating are due to the structure of the CNS sheet and the formation of the TA-Fe layer at the steel/coating interface [54].

In another study, Prasad et al. investigated the inhibitory behavior of *Piper nigrum* leaves (PNL) extract for 410 stainless steel in 0.5 M H_2SO_4 using electrochemical methods. Based on the results of potentiodynamic polarization, in the presence of 500 mg/L of PNL extract, the inhibition efficiency is 94.20% [55].

During another research work, Haldhar et al. investigated the inhibitory behavior of *Solanum surattense* (*S. surattense*) extract for low carbon steel (LCS) in 0.5 MH_2SO_4 using Weight-loss technique and electrochemical methods (Tafel and EIS). Based on the results of potentiodynamic polarization, in the presence of 500 mg/L of *S. surattense* extract, the inhibition efficiency was 93%. SEM and AFM graphical results were used to detect the anticorrosion protective layer outside the alloy. FT-IR results showed the presence of different functional groups and different atoms other than carbon and nitrogen in the protective *S. surattense* extract. Based on the Langmuir adsorption isotherm, the molecules of *S. surattense* extract react with the LCS surface and act as mixed-type inhibitors. DFT data showed high EHOMO and low ELUMO values for the active plant compounds in *S. surattense* extract, which confirmed the inhibitory properties of the extract [56].

During another research work, Halambek et al. investigated the inhibitory behavior of pectin from tomato peel waste (TPP) for Tin in 2% NaCl and 1% acetic acid using weight loss technique and electrochemical methods (Tafel and EIS). Based on the gravimetric results, the highest inhibitory effect was 73.9% at a concentration of 20 g/L of TPP. The electrochemical results showed that due to the tendency of pectins to form a viscous solution, the maximum inhibition efficiency of 60.05–65.5% is obtained at a concentration of 4 g/L. The results of potentiodynamic polarization showed that pectins provide cathodic protection of the tin surface [57].

Wu et al. investigated the inhibitory behavior of green husk extract (WGHE) for magnesium alloys in 0.05 wt% NaCl using electrochemical methods. According to the electrochemical results, the inhibition efficiency was only 44.8% in the presence of 1.0 g/L of WGHE. On the other hand, inhibition efficiency cannot be further improved by increasing the concentration of WGHE. If the magnesium alloy samples were immersed in a 1.0 g/L solution of WGHE for 48 h, inhibition efficiency increased to 92.5%. Based on the results, menadione is one of the main compounds in WGHE that cause corrosion inhibition properties in the extract [58].

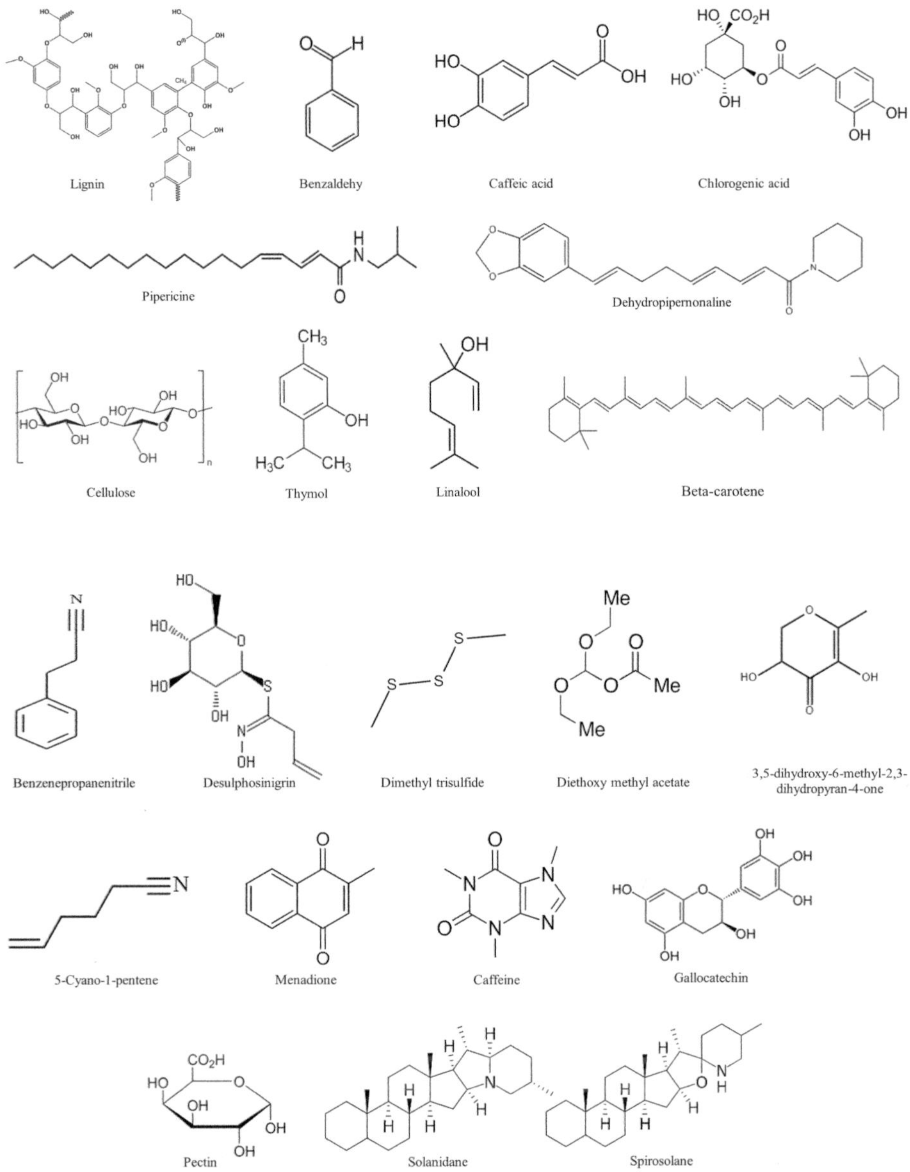

FIGURE 3.4 The active ingredients in waste of plants applied as corrosion inhibitors.

Liu et al. investigated the inhibitory behavior of waste *Platanus acerifolia* leaves (PAL) for steel in 0.06 M chloride using electrochemical methods. Based on the results of potentiodynamic polarization, PAL extract was reported as a mixed-type inhibitor, and as a result, it had the ability to inhibit both cathodic and anodic reactions [59].

Song et al. also investigated the inhibitory property of PAL waste extract for steel in 0.05 M NaCl using electrochemical methods, and it was found that in the presence of 3% of PAL extracted by ultrasound-assisted enzymatic, the IE was 96.65% [60].

Maurya et al. studied *Solanum melongena* steam (brinjal cap) aqueous extract (SMSAE) as anticorrosion substance toward low-carbon steel in 1 M NaCl. An inhibitory efficiency of >70% was obtained in the presence of 500 mg/L of SMSAE using electrochemical techniques [61].

3.14 LEAVES

Wang and colleague studied the corrosion inhibition performance of *Artemisia argyi* leaf extract for carbon steel, and it was shown that the presence of quercetin, jaceosdin, euqatilin, and 2-benzylidenesuccinic acid makes the extract a powerful, eco-friendly corrosion inhibitor candidate [62]. Dehghani et al. have used *Aloysia citrodora* leaf extract to investigate corrosion mitigation on mild steel, and the result showed that by the protonation of oxygen atoms of the *Aloysia citrodora* extract, major component (cinerolone, benzyl alcohol, geranic acid, and geranyl acetate) phytochemical binding on metal surface occurs [63]. Abdelaziz et al. explained the corrosion retardation of *Arbutus unedo L.* leaf extract using experimental and computational approaches [64]. In another study, the water extract of *Artocarpus heterophyllus Lam* leaves was investigated. Electrochemical impedance spectroscopy data indicated anticorrosion efficiency of 97.3% for copper surface in H_2SO_4 environment [65]. Flavonoid, phenol, tannin, psoralen, and bergapten are major Golpar leaf compounds, making the extract act as green corrosion inhibitor with resistance of 6.8, 8.1, 9.2, and 9.8 KΩ cm^2 in 400, 600, 800, and 1,000 ppm extract concentration [66]. In another experiment, Xu et al. explained *Luffa cylindrica roem* leaf extract inhibition mechanism [67]. *Artabotrys odoratissimus* has inhibitory impact on mild steel corrosion in an acidic medium, with an efficiency of 93.27% for 1.25 g/L extract concentration characterized by polarization experiments [68]. Electrochemical study, surface analysis, and quantum chemical calculations were employed to estimate *Crataegus oxyacantha* leaf extract corrosion protection of carbon steel in 1 M HCl, and the maximum inhibitory efficiency was found to be 94% in 3 g/L extract solution [69]. In another research, *Passiflora edulia Sims* leaf extract showed high anticorrosion property by chemically adsorbing on Cu interface and IE of 96% in 800 mg/L extract concentration [70]. The leaf extracts of *Allamanda cathartica* [71], banana [72], *Azadirachta indica* [73], *Aquilaria subintergra* [74], *Andrographis echioides* [75], *Amorphophallus paeoniifolius* [76], and *Citrus aurantium* [77] have corrosion retardation effect. An investigation on *Centipeda minima* leaf extract's anticorrosion effect was conducted by Wang et al., and the main components responsible of this were found to be ethyl 3–(3,4–dihydroxyphenyl) acrylate (EDA), 3',4',5,7–tetrahydroxy–3–methoxy flavone (THM), 1,2–bis(3,5–dimethoxyphenyl) ethylene (BDE), and apigenin (AGN) with IE 96.2%, with extract concentration of 500 mg/L [78]. *Cinnamoum tamala* leaf extract is a powerful corrosion bio-inhibitor with IE of 96.76% at 100 mg/L critical concentration, which means it could be an efficient industrial corrosion inhibitor [79]. Bhawsar et al. performed experimental and computational studies to estimate *Nicotiana tabacum* leaf extract inhibitory impact. The mechanism of action was found to be due to film formation upon metal surface, and inhibition efficiency and surface coverage increase with the increase in extract concentration [80]. *Diospyros kaki* (persimmon) leaves showed IE of 91% on st37 steel in 0.1 M HCl system. Found from electrochemical results, *Diospyros kaki* acts as cathodic-type inhibitor and prevents the reduction of hydrogen ions in the cathodic zone [81]. Mild steel can be protected from corrosion using leaf extract of *Garcinia livingstonei* [82], *Paederia foetida* [83], *Ficus hispida* [84], *Dolichandra unguis-cati* [85], and synthetic hybrid pigment based on zinc acetate–*Cichorium intybus L* leaf extract [86]. *Bambusa arundinacea* leaf extract is another green sustainable corrosion retardant for steel-reinforced concrete [87]. Acrylic resin coating can be protected from cession using Henna leaf extract [88]. To restrain the corrosion of X70 steel in acidic solution, Zuo and colleagues used *Lilium brownii* leaf extract, and it was proved that the extract acts as a mixed-type corrosion inhibitor with physical and chemical adsorption [89]. Jokar et al. investigated *Morus alba pendula* leaf extract's effect as corrosion inhibitor. IE of 93% in 0.4 g/L extract concentration was obtained using electrochemical measurements [90]. Tan et al. performed a study to examine the anticorrosion effect of papaya leaves water extract for Cu in acidic medium. Test results showed the metal surface was perfectly covered due to Cu-S bond and Cu-N formed over metal interfaces [91]. The application of *Psidium guajava Linn* leaf extract as corrosion mitigator of copper in an acidic environment was studied by Wu et al. The four main ingredient responsible for the inhibition effect were found to be (S)-2-amino-5-guanidnopentanoic acid (AGPA), 2–(3,4 dihydroxyphenyl)-3,5,7-trihydroxy-4H-chromen-4-one (DDTC), (S)-2-amino-3-phenylpropanoic acid (APPA), and (2R,3S)-2–(3,4-dihydroxyphenyl) chroman-3,5,7-trial (DPCT) [92]. Khadom et al. studied *Xanthium strumarium* leaf extract as an anticorrosion substance toward low-carbon steel in 1 M HCl. Maximum inhibitory efficiency of 94.82% was achieved using electrochemical techniques [93].

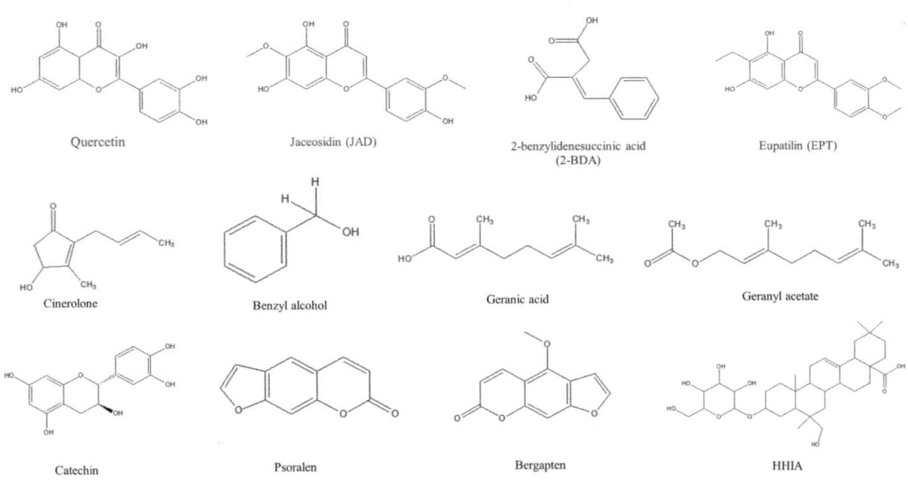

FIGURE 3.5 The active ingredients in leaves of plants applied as corrosion inhibitors.

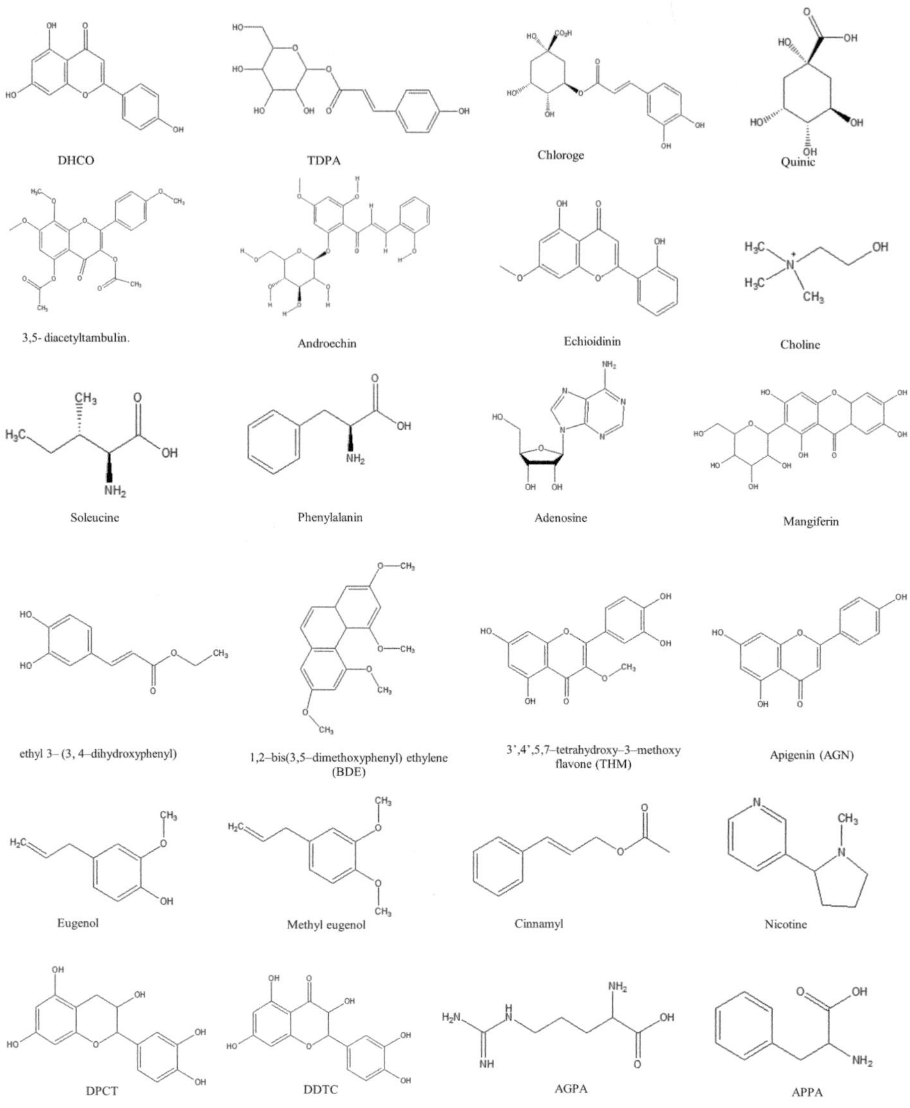

FIGURE 3.5 (Continued)

3.15 STEM

The inhibition of mild steel corrosion in 0.5 M HCl in the presence of *Cnicus benedictus* aerial extract was investigated. *Cnicus benedictus* is a weed from the Asteraceae family. The results suggested that *Cnicus benedictus* extract (CBE) could be a sustainable inhibitor with an efficiency of 92.45% in 0.5 M HCl media and extract concentration of 100 ppm. The inhibition process of CBE in the studied environment can be demonstrated based on molecular adsorption. Ideally, in corrosion investigation, lone pair/π-electrons of the inhibitor molecules form a complex with empty d-orbitals of the metal, and this complex has the ability to protect metal surface. The major phytochemical constituents of CBE are chlorogenic acid and sinapic. These two components have a number of aromatic rings and heteroatoms, which makes them act as Lewis bases. By forming coordinate bonds with the empty d-orbitals of the metal, CBE would be adsorbed on the metal surface and behave as a corrosion inhibitor. During metal standing in acidic solution, the steel surface charge changes to positive as a result of losing electrons and adsorbs Cl^-. At the same time, protonated inhibitor molecules physically adsorb on the metal surface by electrostatic interactions. As discussed earlier, heteroatoms have lone pair of electrons; in the moment, these electrons, and also benzene ring's π-electrons, transfer to unoccupied 3D orbitals of iron atoms and adsorb chemically [94].

In another study, *Corchorus olitorius* stem extract as a corrosion inhibitor for mild steel in 0.5 M H_2SO_4 solution was studied. The results indicated that *Corchorus olitorius* stem extract has inhibitory effect due to the presence of organic constituents, such as glycosides, alkaloids, tannins, flavonoids, and saponins. In this investigation, the highest inhibition efficiency was 93.29%, with inhibitor concentration of 1.0 g/L, temperature of 60°C, and process levels of 4 days [95].

In a study conducted by Haddadi et al., the corrosion-inhibitive effect of zinc cations/*Nepeta pogonosperma* extract was studied for mild steel in a saline solution. *Nepeta pogonosperma*, locally named larzeh (LA), is a plant from the Lamiaceae family. Due to the existence of terpinen-4-ol, α-terpineol, β-pinene, 1,8-cineole, and 4aα-7α-7aα-nepetalactone as major component in LA extract, containing active oxygen functional groups (i.e., C=O and O-H), the extract can act as a chelating agent for zinc cations and so on. The corrosion-retardation performance (CRP) of the *Nepeta pogonosperma* leaf extract can be explained as follows:

> Non-metallic electronegative organic components with oxygen-, nitrogen-, and sulfur-containing functional groups found in LA extract can build dative interactions with Zn^{2+} and occupy D-orbitals of Zn^{2+} cations. When mild steel is submersed in a corrosive solution system, anodically generated Fe^{2+} is ionically exchanged with Zn^{2+} and forms a covalent bond with the iron atom present in mild steel and form an inhibition film over the mild steel surface. Meanwhile, the exchanged Zn^{2+} cations react with Cl- and OH- generated on the cathodic sites and form Simonkolleite. The deposited zinc-based salt and zinc hydroxide components block cathodic sites and reduce cathodic activities over a mild steel surface [96].

Mobin reported that pineapple stem extract (bromelain) can act as a sustainable corrosion inhibitor of low-carbon steel (LCS) in 1 M HCl environment. The inhibitory effect was studied using various techniques as well as electrochemical impedance spectroscopy (EIS), and the result has shown 97.6% inhibition efficiency at bromelain concentration of 1,000 ppm at 338 K. The inhibition process is based on the adsorption of bromelain on the LCS surface caused by the presence of O, N, and S atoms in bromelain. These compounds are protonated when placed in an acidic solution. The steel surface bears a positive charge when corrosion occurs and tends to adsorb Cl^- ions present in the corrosive solution. This makes steel surface charge change from positive to negative and for protonated bromelain components to electrostatically interact with the LCS surface (physisorption). The bromelain molecules

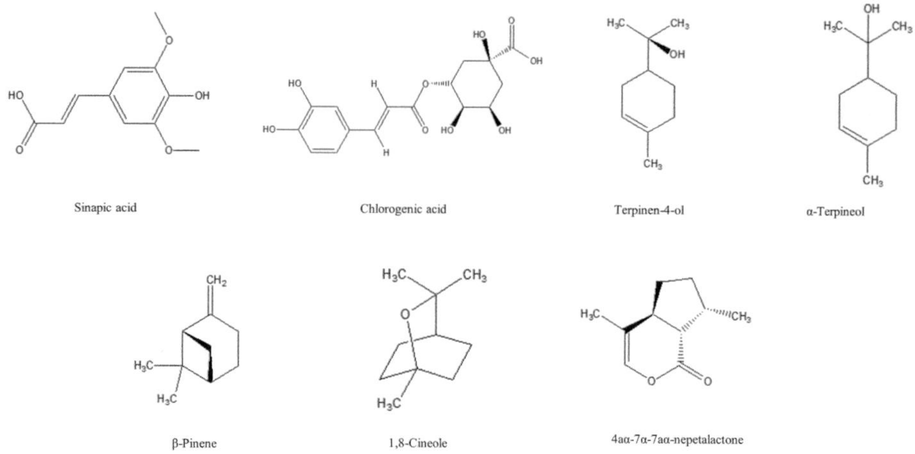

FIGURE 3.6 The active ingredients in the stem of plants applied as corrosion inhibitors.

can also adsorb on the LCS surface by the displacement of water molecules from the LCS surface and sharing of electrons between the heteroatoms and LCS (chemisorption mechanism). Generally, the parameters influencing the adsorption of inhibitor molecule on the LCS surface are hydrogen bonding with N-atom, partial transference of p-electrons, and a lone pair of electrons located on nitrogen, oxygen and sulfur atoms [97].

Hadi et al. found that the stem extract of *Brahmi bacopamonnieri* and *henna* has anticorrosion effect on low-carbon steel in a 0.5 M NaOH solution by avoiding both cathodic and anodic reactions on metal surface. Based on the results, using different concentrations of extract of Brahmi (*Bacopa monnieri*) and Henna solutions ranging between 0.5 and 2% in alkaline solution of 0.5 M NaOH at room temperature and 1 atm, the corrosion inhibitory efficiency increased up to 80% from polarization, while with the weight loss method, the efficiency increased up to 65%. The inhibition happens when extract adsorption on the carbon steel occurs [98].

3.16 GUM

Shahini et al. reviewed experimental and theoretical biopolymer/carbohydrate polymer characteristic as corrosion-inhibitive macro-molecules. The inhibition impact of biopolymers like chitosan, polydopamine, cellulose, and carrageenan, has been widely explored, but the main focus of this literature was to discover lignin, tannin, inulin, extracellular polymeric substance, chitosan, carrageenan, polydopamine, cellulose, and gum as corrosion-inhibitive materials. Several investigating methods were employed to identify their inhibition properties, including electrochemical methods like weight loss, potentiodynamic polarization, EIS and surface analysis like SEM, EDX, XPS, and EDS. The inhibitory mechanism can be explained as the inhibitor molecules adsorbing on the substrate surface and occupying the place of water molecules [99].

Mobin et al. performed a study to determine almond gum (AG) anticorrosion-related properties in 1 M HCl on mild steel (MS). gravimetric, FTIR, NMR, EIS, PDP, SEM/EDAX, UV-visible, DFT, and Monte Carlo simulation were utilized. It was obtained that AG can act as a green, sustainable corrosion inhibitor in acidic solutions in studied temperature (30–60°C), with inhibition efficiency of 96.37% at 300 ppm at 60°C. It was found that inhibition efficiency increases with an increase in temperature and AG concentration. Natural AG components were arabinose, xylose, galactose, and mannose, which make the Fe^{+2} –AG complex on inhibited MS surface and act as a mixed-type inhibitor mostly showing anodic effect [100].

For the determination of *Boswellia serrata* gum (BSG) corrosion mitigation effect, Mobin et al. have used low-carbon steel and an acidic medium. Standard experimental techniques were employed, and the result showed BSG efficient inhibition on lower concentration than other studied plant exudates. The corrosion inhibition process depends on both concentration and temperature, increasing with an increase in BSG concentration and decreasing with temperature. The highest obtained efficiency was 91.9% at 30°C and 500 ppm BSG concentration. The formation of a protective layer is ascribed to physisorption and chemisorption of BSG main polysaccharides (xylose, arabinose, galactose, and glucuronic acid) at the metal surface, however, with physisorption being the dominant one [101].

Ti determine the synergetic corrosion inhibition potential of *Acacia senegal* (Arabic gum) and *Anacardium occidentale* (cashew gum) on mild steel, Arthur has used 1:1 combination of them. The study was conducted in 303 and 333 K temperatures and 0.1 M H_2SO_4. The blend showed good inhibitor property due to the presence of e [6-O-(1-L-maloyl)-alpha-,beta-D-Glcp (veracylglucan A) (22.11%), 6-octadecanoic acid (55.76%), methylhexadecanoate (3.97%), 1-pentadecane carboxylic acid (10.78%), and 9, 12-octadecadienoic acid (7.38%)], identified by GCMS technique [102].

In a study, the application of enzyme- and acid-extracted sugar beet pectin was studied as a natural corrosion inhibitor for mild steel under acidic conditions. Several assessments were used, and the result showed a connection between corrosion efficiency and the pectin extraction method. Enzymatic-extracted ones have higher corrosion than the other. As revealed by polarization curves, both types of pectin are mixed type, with predominant inhibition at the cathodic site [44].

Exudated gum from *Raphia hookeri* was studied as aluminum corrosion inhibitor in 0.1 M HCl media. It was found that at experiment temperature (30–60°C), inhibition efficiency increases with an increase in inhibitor concentration and has diverse relation with temperature. The inhibition effect occurs due to the presence of phytochemical constituents in the exudate, which can spontaneously adsorb on metal surface [103].

Guar gum is a natural polysaccharide that exists in leguminous family tree seeds commonly grown in Pakistan, India, and some parts of North and South America. The extracted guar gam has distinctive properties, such as water-soluble, non-ionic, biocompatible, biodegradable, and having the presence of heteroatoms, making it an excellent biodegradable corrosion inhibitor. *Guar gum* is a galactomannan polysaccharide with a structure that consists of a straight chain of D-mannopyranosyl with (β (1 $\rightarrow$ 4) glycosidic linkage backbone and a side-chain of D-glucopyranosyl with (α (1 $\rightarrow$ 4) glycosidic linkage. Due to its biodegradability, economic feasibility, biocompatibility, and biosafety, guar gum has been widely used in pharmaceutical, biomedical, food industry, and wastewater treatment. With increasing attention to guar gum, extensive research have been conducted [104, 105].

In 2014, Roy et al. performed a study to evaluate the inhibitory effect of polyacrylamide-grafted guar gum (GG-g-PAM) with different grafting percentage on mild steel immersed in an acidic solution. It was proved that grafted copolymer (up to 86% grafting) inhibition efficiency has been maintained higher than 90%, and in higher grafting, inhibition efficiency gradually decreases. GG-g-PAM acts as a mixed-type inhibitor and forms a layer on mild steel surfaces with endocyclic and exocyclic O atoms of GG sugar units and creates a complex with Fe^{2+} [106].

In another study, Singh et al. have synthesized a natural polysaccharide composite consisting of guar gum and ethyl acrylate (GG-EEA). The synthesized composite was introduced to p110 steel in 15% HCl medium. Corrosion technique result showed GG-EEA can act as a mixed-type corrosion inhibitor by physically and chemically adsorbing on active sites of the metal surface and blocking them. GG alone has corrosion efficiency of 77.5%, and that for GG-EEA is 92.3% at 500 mg/L [107].

In another study, the corrosion inhibition ability of guar gum-grafted 2-acrylamido-2-methylpropanesulfonic acid (GG-AMPS) toward carbon steel was examined. The corrosive solution was ASTM standard D1141–98(2013) artificial seawater. Primary corrosion techniques were used, and polarization results indicated GG-AMPS cathodic majority. EIS indicated that with an increase in GG-AMPS concentration, charge transfer resistance rises, and with scanning electron microscopy formation, GG-AMPS film over metal surface was observed [108].

In another investigation, Messali et al. studied guar sum corrosion inhibition efficiency on carbon steel and 2 M H3PO4 solution. At studied concentration (0.1 to 1.0 g/L) and 298e328 K–obtained results from weight loss and electrochemical methods showed by increasing experiment temperature, inhibition efficiency decreased and increased by adding inhibitor up to 1.0 g/L. EIS showed that the treated carbon steel surface polarization resistance (Rp) is significantly higher than for the untreated one [109].

Also, surfactants can be added to guar gum and have synergistic impact on mitigating corrosion, as Shamsheera et al. studied in 2021. They added Tween-80 and Triton-X, two non-ionic surfactants, to GG in n 0.5 M HCl, and the studied metal was mild steel. Results obtained from weight loss and electrochemical methods proved synergistic, remarkable inhibition efficiencies of 97.13% and 95.50% for GG:Tween-80 and GG:Triton-X, respectively [110].

In another research, Biswas et al. studied the Grafting effect of f gum acacia (GA) with polyacrylamide on corrosion inhibition of mild steel in acidic solution. The combination of 1 g gum acacia and 3 g acrylamide showed high corrosion mitigation efficiency of 94.08% at 0.3 g/L and was found to be a mixed-type inhibitor. The adsorption process on metal surface followed Langmuir adsorption isotherm. It was proved by atomic force microscopy that it forms a protective film over a metal surface [111].

In a study to evaluate terebinth extract's ability as green corrosion inhibitor for iron in saline medium, maximum efficiency was found to be 86.4% at 3,000 ppm. The major compounds, found to be a-pinene (32.65–50.58%), limonene (6.88–15.07%), and a-terpineol (2.50–5.15%), resulted from analyzing essential oil by GC/MS. SEM, EDX, and IR were employed to analyze metal surface and implied formation of a protective barrier through metal [112].

In a comparative study on *Ferula assa-foetida* (*F. assa-foetida*) and *Dorema ammoniacum* (*D. ammoniacum*), oleo-gum resin exudate as inhibitor candidate for mild steel corrosion in acidic medium was conducted. Common electrochemical techniques were carried out. The results of potentiodynamic polarization curves confirmed both oleo-gums are mixed-type corrosion inhibitors. The most relevant compounds to their inhibitory activity were umbelliferone, ferulic acid, farnesiferol-C, farnesiferol-B, 1-propenyl propyl disulfide, (R)-2-butyl-1-propenyl disulfide and 1(1-methylthiopropyl) -1-propenyl disulfide in *F. assa-foetida*, and linalool, linalyl acetate, citronellyl acetate, and ammoresinol in *D. ammoniacum* [113].

In another study, xanthan gum and xanthan-gum-graft-poly(acrylamide)(XG-g-PAM) corrosion inhibitors' impact on mild steel in an acidic solution was evaluated. The inhibition mechanism was found to be caused by the creation of a barrier over the surface, and this barrier forms with the aid of donor atom of the molecules (N, O, S, etc.) and surface charge [114].

Azzouzi et al. compared Morocco (GAMo), Mauritania (GAMau), and Senegal (GASe) gum Arabic variants as corrosion mitigator for mild steel in 1 M HCl. Test results confirmed their corrosion ability as mixed-type inhibitor with efficiency of 96%, 95%, and 94% for GAMo, GASe, and GAMau, respectively, in 1 g/L inhibitor concentration [115].

Arukalam et al. reported *Lupinus arboreus* gum extract corrosion inhibitory characteristics in epoxy-based coating. Phytochemical analysis was done to identify bioactive constituents, and then the extract was encapsulated with formaldehyde (UF) and melamine-formaldehyde (MF). EIS result confirmed LA gum extract inhibition effectiveness [116].

Locust bean gum (LBG) can also be an excellent green corrosion inhibitor, as studied by Guo et al. for Q235 steel in 0.5 M H_2SO_4 medium. Obtained from potentiodynamic polarization test in 5 mM (LBG) concentration and 298 K temperature, the corrosion inhibition efficiency was 89.8%. LBG adsorption on metal surface obeyed Langmuir single-layer model [117].

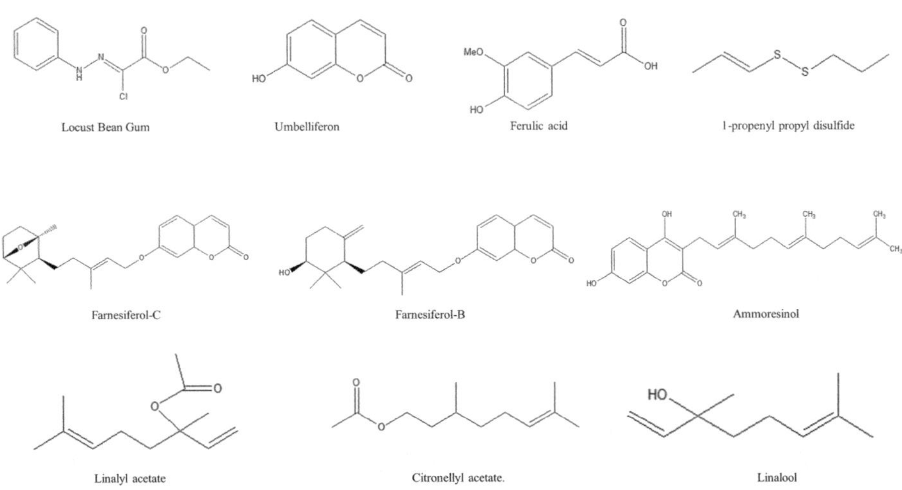

FIGURE 3.7 The active ingredients in the gum of plants applied as corrosion inhibitors.

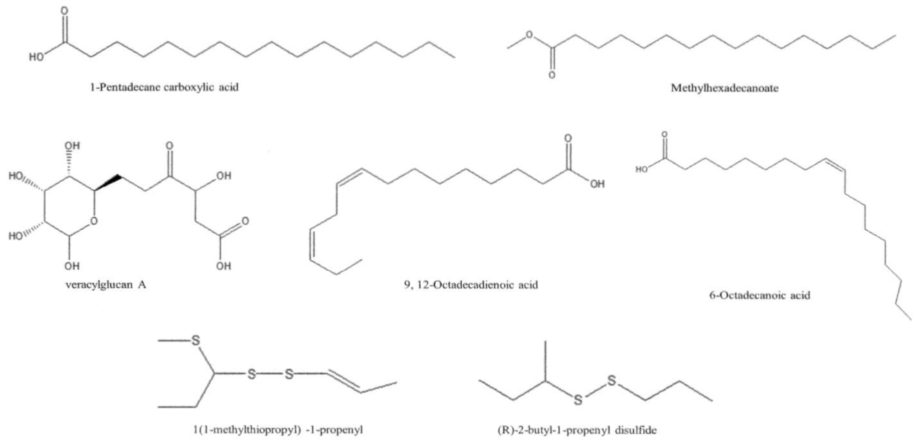

FIGURE 3.7 (Continued)

3.17 FLOWER

Raghavendra and Bhat have studied Areca flower extract in 0.5 M HCl solution as Al corrosion inhibitor using weight loss, impedance, and Tafel plot techniques. Derived from mass loss technique, on the metal and medium solution interface, a strong bond would be formed with the aid of Areca flower extract molecules following Langmuir adsorption model. Tafel plot indicated that Areca flower extract molecules can replace the H_2O molecules and adsorb on the active Al sites, limiting dissolution reaction by interacting with cathodic and anodic sites, and protect Al metal from corrosion. Also, the EIS result has shown that introducing Areca flower extract to the metal results in higher semicircle (Nyquist plots) in comparison with bare system [118].

Khalid Ahmed et al. investigated the corrosion inhibitor efficiency of bee pollen extract (BPE) for pure copper in an acidic solution. Gallic acid, vanillic acid, rutin, D-gluconic acid, linoleic acid ethyl ester, stearic acid, behenic acid, oleanolic acid, propanoic acid, myristic acid, benzenepropanoic acid, 9,12,15-octadecatrienoic acid, luteolin, kaempferol, and palmitic acid are BPE identified from gas chromatography mass spectrometry. Mitigating corrosion efficiency can be attributed due to their existence in BPE. Experimental techniques such as weight loss, electrochemical impedance spectroscopy, and potentiodynamic polarization tests were employed to determine the corrosion properties of pure copper immersed in 1 M HCl with BPE. It was found that the adsorption process follows Langmuir-type isotherm, with maximum inhibition efficiency of 94.5% with 7 g/L BPE concentration [119].

In another study conducted by Nathiya et al., corrosion inhibition ability of Borassus flabellifer coir dust for aluminum metal in 1 M H_2SO_4 was examined. The Borassus flabellifer coir dust extract was extracted by water (BFDWE) and methanol (BFDME). The results of electrochemical, non-electrochemical methods indicated that in a concentration of 0.4 g/L BFDME and BFDWE, ~66.8% and

~52% inhibition efficiency was observed, respectively, due to the physisorption of molecules of BFDME and BFDWE on top of metal surface. Also, it was found that BFDME and BFDWE act as s mixed-type inhibitors with significant cathodic characteristic [120].

Sedik et al. studied the corrosion inhabitation of dardagan fruit (DF) grown in Bingöl, Turkey. Mild steel in an acidic solution in contact with DF was studied using electrochemical techniques. With the aid of spectroscopic studies, fumaric, vanillic gentisic, and chlorogenic acids were found to be major components in DF extract and make DF extract a promising green corrosion inhibitor. The electrochemical results have shown 92% corrosion inhibition efficiency after 1 hour, and 97% after 6 hours, immersion to 3,000 ppm DF extract containing 1 M HCl solution. This means protective film formation over mild steel has diverse connection with exposure time and DF concentration [121].

Mechbal et al. have used the aqueous extract of the leaves, flowers, and stems of *Cistus monspeliensis L* (CMAE) to study antioxidant and anticorrosive characteristics on mild steel in a hydrochloric acid solution using electrochemical methods. The highest inhibition efficiencies of 88%, 81%, and 70% for 0.2 g/L CMAE were observed on leaves, flowers, and stems, respectively. These data show the inverse relationship between concentration and inhibitory efficiency. The CMAE adsorption on the substrate surface follows the Langmuir adsorption isotherm model [122].

Tabatabaei Majd et al. studies a novel natural corrosion mitigation of mild steel immersed in saline solution with the presence of petal extracts of the primrose flower (PPE) and zinc cations. The inhibition efficiencies for PPE and Zn alone were found to be between 40 and 80%, while the combination of 300 ppm PPE + 700 ppm Zn showed 95.3% protection degree. The inhibition mechanism can be attributed to PPE:Zn containing solutions/actions derived from potentiodynamic polarization spectroscopy (PPS) and electrochemical impedance spectroscopy (EIS). The complex formation between the organic molecules of the PPE and Fe2+/Zn^{2+} cations protects mild steel surface [123].

In another study, Tabatabaei Majd and colleagues studied the effect of *Primula vulgaris* flower aqueous extract (PVFE) as a mild steel corrosion inhibitor exposed to an acidic medium. Electrochemical measurements illustrated in the sample a 95.51% inhibition efficiency in a solution of 1,000 ppm PVFE, and atomic force microscope proved a film formation on the mild steel dipped in PVFE. This film formation is caused by catechin, cinnamic acid, protocatechuic acid, and rutin compounds in PVFE which are protonated. When the protonated PVFE molecules interact with OH- solution ions, they become deprotonated and adsorbed on the metal surface [124].

Molaeipour et al. studied the application of *Stachys byzantina* extract as a green bioactive corrosion inhibitor for carbon steel. They aimed to create a long-lasting protective film on carbon steel surface with the interaction of SB and zinc cations. Electrochemical measurement methods were used to identify inhibitory efficiency. After a day of immersing steel in a solution of 700 ppm SB and 300 ppm of ZN, corrosion inhibition behavior was about 92%, obtained from EIS. Potentiodynamic polarization has also shown a significant reduction of the i_{corr} [125].

In a study conducted by Furtado et al., wine industry residues were examined as microencapsulated corrosion inhibitor. The examination was held on API P110 steel in 1 M HCl. The result of gravimetric and electrochemical experiments proved weight loss efficiencies of 83.81% at 303 K with 4.0 g/L of the extract even after storage. Derived from polarization tests, the inhibitor adsorbs on both the cathodic and anodic regions [126].

Mourya et al. examined the corrosion inhibition of *Tagetes erecta* (marigold flower) (TEE) extract for mild steel in 0.5 M H_2SO_4 solution. Gravimetric, electrochemical impedance, spectroscopic, and potentiodynamic polarization measurements were used and revealed that TEE acts as a mixed-type inhibitor, and the adsorption of TEE forms a monolayer on the metal surface. The adsorption process is physically caused by lutein, TEE's major component [127].

Haque et al. studied the hydrochloric acid–induced corrosion of mild steel using *Thevetia peruviana* (Kaner) flower extract (TPFE). Electrochemical studies pointed out that TPFE acts as a mixed- and interface-type of corrosion inhibitor with efficiency of 91.24% at 200 mg/L concentration. Density functional theory (DFT) and molecular dynamics (MD) expressed that TPFE interacts with donor–acceptor interactions, and its components form a layer over mild steel surface and acquire flat or horizontal orientations. Major constituents present in TPFE extract were found to be cannogenin, connonigenol, digitoxigenin, and thevetin B [128].

Divya et al. performed a study to examine *Tithonia diversifolia* flower's (TDF) corrosion inhibition effect for mild steel in 1 M HCl using conventional mass loss, electrochemical impedance, and potentiodynamic polarization techniques. Inhibition efficiency was found to be 94.39% and 90.1% at temperatures of up to 325 K, increasing with high concentrations of TDF. Concluded from potentiodynamic polarization curves, TDF acted as a mixed inhibitor in an acidic medium. Impedance measurements inferred a change in the charge transfer resistance and double layer capacitance, implying adsorption of TDF on the metal surface. FT-IR and optical electron microscope were employed for surface analysis [129].

Rajamohan et al. used an eco-friendly biomass from *Ziziphus spina-christi* extract to investigate its corrosion protection effect on carbon steel under acidic conditions. Using mass loss technique, the extract exhibited high weight loss efficiencies (91.35%) at 300.15 K at 810 ppm inhibitor concentration. Corrosion rate increased with the dilution of extract solution, and surface coverage enhanced. Electrochemical analysis proved *Ziziphus spina-christi* extract is a mixed-type inhibitor and the increase in corrosion resistance through double-layer capacitance [130].

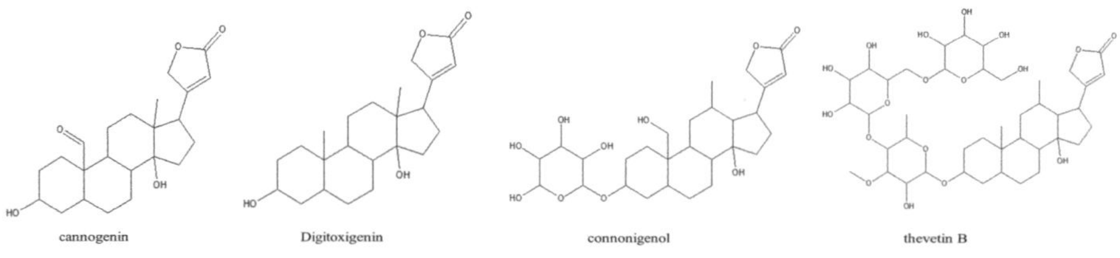

FIGURE 3.8 The active ingredients in the flower of plants applied as corrosion inhibitors.

TABLE 3.1
The Detail of Plant Extracts as a Corrosion Inhibitor

Num.	Plant Extract	Anticorrosion Active Constituents	System	Techniques Used	Highest Protection Efficiency (%)	Inhibitor Concentration	Reference
			Fruit				
1	*Citrullus lanatus* fruit	Citrulline, hesperetin, resveratrol, and riboflavin	Mild steel in 1 MHCl	Electrochemical (EIS and polarization) and surface morphological (SEM, AFM, contact angle) techniques	91%	800 ppm	[3]
2	*Cupressus arizonica* fruit essential oil	a-pinene, myrcene, and limonene	Carbon steel (CS) in 1 M HCl	EIS, polarization, and SEM	93%	0.5 g/L	[4]
3	*Garcinia cambogia* fruit rind	Cyanidin, garcinol, hydroxy citric acid, malic acid, ascorbic acid, and tartaric acid	Mild steel in 0.5 N HCl	EIS, polarization, and SEM	91.50% 92.85%	4%V/V	[5]
4	*Garcinia indica* fruit peel	Cyanidin 3-glucoside, cyanidin sambubioside, and hydroxy citric acid	Mild steel in 1 M HCl	Weight loss, EIS, polarization, and SEM	90%	4%V/V	[6]
5	Waste of lychee fruit	(-)-epicatechin, (-)-epigallocatechin, and cyanidin-3-glucoside	Mild steel in 0.5 M HCl	Weight loss, EIS, polarization, and SEM	97.95%	600 mg/L	[7]
			Leaf				
6	*Pimenta dioica* leaf	Eugenol, methyl eugenol, β-caryophyllene, myrcene, limonene, 1,8-cineole	Mild steel in 0.5 M and 1 M HCl	EIS, polarization, and SEM	95.00%	4%V/V	[8]
7	*Chromolaena odorata* leaf	–	Mild steel in 1 M HCl	EIS, polarization, SEM, and HRTEM	92.39%	1.0 g/L	[9]
8	*Dryopteris cochleata* leaf	–	Mild steel in 1 M H_2SO_4	Weight loss, EIS, polarization, and SEM	95.09%	2,400 ppm	[10]

(Continued)

TABLE 3.1 (*Continued*)
The Detail of Plant Extracts as a Corrosion Inhibitor

Num.	Plant Extract	Anticorrosion Active Constituents	System	Techniques Used	Highest Protection Efficiency (%)	Inhibitor Concentration	Reference
9	Green eucalyptus leaf	–	Mild steel in 1 M HCl	EIS, polarization, and SEM	88%	000 ppm	[11]
10	Ginkgo leaf	Isorhamnetin, sciadopitysin, 6-hydroxykynurenic acid, and 4-O-methylpyridoxine	X70 steel in 1 M HCl	EIS, polarization, and SEM	92.50%	200 mg/L	[12]
11	*Lagerstroemia speciosa* leaf	–	Mild steel in 1 M HCl	Weight loss, EIS, polarization, SEM, and density functional theory (DFT)	94%	500 ppm	[13]
12	*Lannea coromandelica* leaf	–	Mild steel in 1 M H_2SO_4	Weight loss, EIS, polarization, and SEM	93.8%	200 mg/L	[14]
13	Olive leaf	Oleuropein	Pure copper in 0.5 M NaCl	EIS, polarization, and SEM	90.00%	2.42 mmol L^{-1}	[15]
14	*Syzygium samarangense* leaf	DMD, PDH, TCT, and DTT	Q235 steel in in 0.5 M H_2SO_4	Electrochemical ways, topography research, and theoretical calculations	90.00%	400 mg/L	[16]
15	Tropical almond (*Terminalia catappa*) leaf	Gallic acid, geranine, quebulagic acid, and galagic acid	Zinc and carbon steel 1020 (according to ASTM standards for gravimetric testing)	Gravimetric test	78.41%	–	[17]
16	*Syzygium cumini* leaf	–	Carbon steel in 0.5 M H_2SO_4	PDP, EIS, and SEM	83.5%	0.3 g/L	[18]

17	Olive leaf	Diisooctyl phthalate, Hexanedioic acid, mono(2-ethylhexyl) ester, 9-octadecenamide, (Z)-, oxiraneoctanoicacid, 3-octyl-, methyl ester, cis, oleic acid, 1-eicosene, 1,2-benzenedicarboxylic acid, butyl 8-methylnonyl ester, n-hexadecanoicacid, benzene, (1-methyldodecyl)-, phenethylamine, 3-benzyloxy-2-fluoro-b-hydroxy-, 1-octadecanol, phenol,2,6-bis(1,1-dimethylethyl)-4-ethyl-, tetradecanoicacid, 1,7-di-iso-propylnaphthalene, cyclopentaneacetic acid,3-oxo-2-pentyl-, methyl ester, and benzophenone	Mild steel in 0.1 M NaOH + 0.5 M NaCl	PDP, EIS, and Mott–Schottky	91.9%	–	[19]
18	*Malva sylvestris* leaf	–	Mild steel in 2.0 M HCl	Weight loss, PDP, EIS, and SEM	93%	2.0 g/L	[20]
			Seed				
19	*Retama monosperma* (L.) Boiss. seed	N-methylcytisine, dehydro-cytisine and cytisine	Carbon steel in 1 M HCl	Electrochemical and surface characterization techniques	94.4%	400 mg/L	[21]
20	Pehuén (*Araucaria araucana*) seed	Gallic acid, catechin, quercetin, and tannin	Steel in 1 M HCl	Weight loss	61%	400 μg/mL of HCl	[22]
21	*Cannabis sativa* L. seed	Linoleic acid, oleic acid, α-linolenic acid, and palmitic acid	E24 carbon steel in 1 M HCl	Weight loss and electrochemical methods	92.4%	1 g/L	[23]
22	*Ceratonia siliqua* L seed	Vanillin, α-tumerone, 2-heptanal, mome-inositol, 1-dotiacontanol, and linoleic acid	Carbon steel in 1 M HCl	Open circuit potential, polarization curves, impedance spectroscopy, and both thermodynamic and kinetic parameters (ΔH, ΔS, and Ea)	95%	100 mg/L	[24]

(*Continued*)

TABLE 3.1 (*Continued*)
The Detail of Plant Extracts as a Corrosion Inhibitor

Num.	Plant Extract	Anticorrosion Active Constituents	System	Techniques Used	Highest Protection Efficiency (%)	Inhibitor Concentration	Reference
23	Clove seed	Campesterol, eugenitin, geraniol, maslinic acid, methyl salicylate, and rhamnetin	Mild steel in 1 M HCl	Electrochemical combined with surface morphological, and theoretical (molecular simulation connected with density functional theory) techniques	93%	800 ppm	[25]
24	Esfand seed	–	Carbon steel in 1 M HCl	EIS, polarization, SEM, Monte Carlo (MC), molecular dynamics (MD), and quantum mechanics	98.8%	300 ppm	[26]
25	*Gloriosa superba* seed	Colchicine and colchicoside	Carbon steel in 1 M H_2SO_4	Weight loss, EIS, polarization, and SEM	93.84%	700 mg/L	[27]
26	*Piper guineense* seed	–	Q235 carbon steel in 0.25 M H_2SO_4 + 0.5 M NaCl	Weight loss, EIS, and SEM	> 98%	0.5 g/L	[28]
27	Grape seed	–	Q235 steel in seawater	Weight loss, EIS, and SEM	96.06%	200 mg/L	[29]
28	*Heracleum persicum* seed	–	Mild steel in 1 M HCl	Weight loss, EIS, polarization, and SEM	96.00%	400 ppm	[30]
29	Lemon seed	–	Mild steel in 1 M HCl	Electrochemical studies (EIS and polarization), molecular dynamics simulations, and quantum chemical calculations	~98%	–	[31]
30	Mustard seed	–	Mild steel in 1 M HCl	Weight loss, EIS, polarization, and SEM	97%	200 mg/L	[32]

31	*Ocimum basilicum* seed	–	Mild steel in 1 M HCl	EIS, polarization, and SEM	95.12%	1.0 g/L	[33]
32	*Peganum harmala* seed	Harmaline, harmalol, harmane, harmol, vasicine, and vasicinone	Mild steel in 1 M HCl	Electrochemical methods (EIS and polarization), surface analyses (AFM, SEM, and contact angle), and theoretical studies (DFT, MC, and MD)	>95%	800 ppm	[34]
33	*Phyllanthus emblica* seed	Syringaldehyde, methyl caffeate, penicillide, and 3,4,8,9,10-pentahydroxydibenzo [b, d] pyran-6-one	410 stainless steel in 15% M HCl	Weight loss, EIS, polarization, and SEM	92.43%	4 g/L	[35]
34	Quince seed	Alanine, glycine, leucine, ascorbic acid, and n-O-caffeoyl quince acid	Mild steel in 1 M HCl	EIS, polarization, and SEM theoretical studies (DFT and MD)	95%	800 ppm	[36]
35	Quinoa seed	–	Carbon steel in 1 M HCl	Electrochemical tests, including Tafel polarization, EIS, and SEM	98%	4 g/L	[37]
36	*Retama monosperma* (L.) Boiss. seed	Cytisine derivatives (cytisine, dehydrocytisine and N-methyl-cytisine)	Mild steel in 1 M HCl	Quantum and dynamic investigations (DFT, MC, and MD)	–	–	[38]
37	*Ocimum basilicium* seed	–	2024 -T3 aluminum alloy in 3 wt% NaCl	Weight loss, EIS, polarization, and SEM	95.5%	1 g/L	[39]
38	Pumpkin seed	9,12-octadecadienoic acid (Z, Z)-, palmitic acid, linoleic acid, and ethyl (9Z, 12Z) -9,12-octadecadienoate	7075-T6 aluminum alloy in 3.5% NaCl	EIS, polarization, and SEM	95%	1 g/L	[40]
39	*Opuntia dillenii* seed	Linoleic acid	Iron in acid rain	Weight loss, EIS, polarization, and SEM	99.19%	1,000 ppm	[41]

(*Continued*)

TABLE 3.1 (*Continued*)
The Detail of Plant Extracts as a Corrosion Inhibitor

Num.	Plant Extract	Anticorrosion Active Constituents	System	Techniques Used	Highest Protection Efficiency (%)	Inhibitor Concentration	Reference
40	Passion fruit seed oil microemulsion	–	P110 carbon steel in CO_2–saturated brine	Weight loss, linear polarization resistance, electrochemical impedance spectroscopy, and polarization curves	>99%	1.5%V/V	[42]
			Waste				
41	Turnip peel	Dimethyl trisulfide, 5-cyano-1-pentene, diethoxy methyl acetate, 3,5-dihydroxy-6-methyl-2,3-dihydropyran-4-one, desulphosinigrin, and benzene propane nitrile	Copper in 3.5% NaCl	Tafel polarization, EIS, and SEM	91.0%	20%V/V	[43]
42	Sugar beet pulp pectin	Galacturonic acid, arabinose, galactose rhamnose, glucose, and ferulic acid	Mild steel in 1 M HCl	Weight loss, EIS, SEM, K_{ads}, and ΔG_{ads}	93.9%	1 g/L	[44]
43	Hard shell of pistachio nut	Cellulose and lignin	Mild steel in 1 M HCl	EIS, polarization methods, SEM, DFT, and MD	92%	800 ppm	[45]
44	*Akebia trifoliate* Koidz peel	–	Mild steel in 1 M HCl	EIS, polarization methods, SEM, and MD	>90%	800 mg/L	[46]
45	Almond waste	–	C38 carbon steel in HCl	Gravimetric analyses, electrochemical tests, and surface characterization	92.76%	0.5 g/L	[47]
46	Carrot pomace	Pectin, cellulose, lignin, and beta-carotene	Mild steel in 1 M HCl	Weight loss, EIS, polarization, SEM, and theoretical studies (DFT and MD)	97%	400 ppm	[48]
47	*Citrus limetta* pulp waste	Linalool	410 stainless steel in 15% HCl	Weight loss, EIS, polarization, SEM, and theoretical studies (DFT, MC, and MD)	90.90%	4 g/L	[49]

48	Tea factory waste	Caffeine, catechin, and gallocatechin	Boiler-quality (BQ) steel in 1 (M) HCl	PDP, EIS, and SEM	90.28%	500 mg/L	[50]
49	Waste rice husk	Cellulose	Mild steel in 1 M HCl	Machine learning, EIS, and SEM	97.73%	80 ppm	[51]
50	Palm kernel shell	–	Stainless steel in 0.5 M H_2SO_4	Gravimetric, electrochemical tests (polarization methods), and SEM	93%	4 g	[52]
51	Apricot pomace	Benzaldehyde, gallic acid, caffeic acid, chlorogenic acid, thymol, linalool, β-cyclocitral, catechin, and terpineols	Mild steel in 0.5 M NaCl	Weight loss and SEM	89.06%	0.5 mg/L	[53]
52	Waste biomass (cornstalks)	Tannic acid	Steel in 3.5 wt% NaCl	EIS, polarization methods, and SEM	–	–	[54]
53	Eco-waste (*Piper nigrum* leaves)	Isopiperolein, dehydropipernonaline, and pipericine	410 stainless steel in 0.5 M H_2SO_4	EIS, polarization, SEM, and theoretical studies (DFT and MD)	94.20%	500 mg/L	[55]
54	Biomass plant waste (*Solanum surattense*)	Solanidane, spirosolane, and avenacoside	Low-carbon steel (LCS) in 0.5 M H_2SO_4	Weight loss technique, electrochemical methods (Tafel and EIS), SEM, and DFT	93%	500 mg/L	[56]
55	Tomato peel waste	Pectin	Tin in 2% NaCl and 1% acetic acid	Weight loss, EIS, and polarization	73.9%	20 g/L	[57]
56	Walnut green husk	Menadione	Magnesium alloys in 0.05 wt% NaCl	Tafel, EIS, SEM, and MD	92.5%	1.0 g/L	[58]
57	Waste *Platanus acerifolia* leaves	5,8-dihydroxy-2–(2-phenylethyl) chromone, astragalin, buddlenoid A, cyclo-(phe-phe), cyclo-(Tyr-Leu), docosanoic acid, hyperosid, licochalcone A, pinocembrin, procyanidin B1	Steel in 0.06 M chloride	PDP, EIS, and SEM	–	–	[59]

(*Continued*)

TABLE 3.1 (*Continued*)
The Detail of Plant Extracts as a Corrosion Inhibitor

Num.	Plant Extract	Anticorrosion Active Constituents	System	Techniques Used	Highest Protection Efficiency (%)	Inhibitor Concentration	Reference
58	Waste *Platanus acerifolia* leaves	Quercetin, (-)- epicatechin, 2' 6'-dihyaroxy-4, 4'-dimethoxy dihydrochalcone, procyanidin B1, quercetin-3-β-d-glucoside, betaine, 5' 7' -dihydroxychromone, gibberellin A7	Steel in 0.05 M NaCl	PDP, EIS, and SEM	96.65%	3%	[60]
59	Waste *Solanum melongena* stem (brinjal cap)	–	Low-carbon steel in 1 M NaCl	Weight loss, Tafel, EIS, and SEM	>70%	500 mg/L	[61]
			Leaves				
60	*Artemisia argyi*	Quercetin (QUE), jaccceeosidin (JAD), 2-benyldenesuccinic acid (2-BDA), eupatilin (EPT)	Carbon steel in HCl solution	Tafel, EIS, SEM, and MD	96.40%	–	[62]
61	*Aloysia citrodora*	Cinerolone, benzyl alcohol, gerenic acid, geranyl accetate	Mild steel (MS) inhibitor against 1 M HCl	EIS, polarization, SEM, and theoretical studies (DFT, MC, and MD)	94%	–	[63]
62	*Arbutus unedo* L.	Quercetin, catechin	Mild steel in HCl medium	Electrochemical methods (Tafel and EIS)	91.72%	0.5 g/L	[64]
63	*Artocarpus heterophyllus* Lam	2–(2,4-dihydroxyphenyl)-5,7-dihydroxy-4H-chromen-4-one (DPD), (2R,3R)-2–(2,4-dihydroxyphenyl)-3,5,7-trihydroxy-chroman-4-one (DPT), 3–(4-hydroxyphenyl) acrylic acid (HPT), and 5–(6-hydroxybenzo-furan-2-yl) benzene-1,3-diol (HBB)	Copper in H_2SO_4 environment	Electrochemical methods and theoretical studies (QCC and MD)	97.30%	500 ppm	[65]

64	Golpar	Psoralen, phenol, flavonoid, begapten, tannin	Mild steel (MS) HCl	Electrochemical methods (Tafel and EIS)	91%	1,000 ppm	[66]
65	*Luffa cylindrica Roem.*	(4S, 6S, 6R, 10S, 12R, 14R)-9-(hydroxymethyl)-2,2,6,9-hexamethyl-10-((2R,3R,4S, 5S,6R)-3,4,5-trihydroxy-6-(hydroxymethyl)-tetrahydro-2H-pyran-2-yloxy)-icosahydropicenne-4a-carboxylic acid (HHIA), 5,7-dihydroxy-2–(4 -hydroxyphenyl)-4H-chromen-4-one (DHCO), and (2S,3R,4S,5S,6R)-3,4,5-trihydroxy-6-(hydroxymethyl)tetrahydro-2H-pyran-2-yl 3–(4-hydroxyphenyl)acrylate (TDPA)	Cu metal in H_2SO_4 environment	Electrochemical methods and theoretical studies (MD)	91.3%	600 ppm	[67]
66	*Artabotrys odoratissimus*	Flavonol glycosides like artabotryside derivatives; flavonoids such as taxifolin, 7-O-glucoluteolin, and apigenin glycoside; and two organic acids, fumaric and succinic acid; structures of quercetin and kaempferol	Mild steel in 0.5 M H_2SO_4 media	Weight loss, Tafel, EIS, and SEM	93.27%	1.25 g/L	[68]
67	*Crataegus oxyacantha*	Quercetin, chlorogenic acid, quinic acid, catechin	Carbon steel in 1 M HCl	Weight loss, Tafel, EIS, and SEM	94%	3 g/L	[69]
68	*Passiflora edulia* Sims	PTH, DMT, BHT	Copper in sulfuric acid solution	Tafel, EIS, and SEM	96%	800 mg/L	[70]
69	*Allamanda cathartica*	Ursolic acid, pinitol, betulinic acid, 9,12,15-octadecatrioinc acid	Mild steel in 1.0 M H_2SO_4	Tafel, EIS, and SEM	–	–	[71]
70	*Amorphophallus paeoniifolius*	3,5-diacetyltambulin	Mild steel surface in 1 M hydrochloric acid	Weight loss technique, electrochemical methods (Tafel and EIS), and SEM	92.49%	10%v/v	[76]
71	*Andrographis echioides*	Androechin and echioidinin	Mild steel in 1.0 M HCl	Weight loss technique and electrochemical methods (Tafel and EIS)	79.48%	–	[75]

(Continued)

TABLE 3.1 (*Continued*)
The Detail of Plant Extracts as a Corrosion Inhibitor

Num.	Plant Extract	Anticorrosion Active Constituents	System	Techniques Used	Highest Protection Efficiency (%)	Inhibitor Concentration	Reference
72	*Aquilaria subintergra*	Choline, isoleucine, adenosine, phenylalanine, mangiferin	Mild steel in 1 M HCl	Weight loss technique, electrochemical methods (Tafel and EIS), and SEM	93%	1,500 ppm	[74]
73	*Azadirachta indica*	N.M	Copper corrosion in blue water	Weight loss technique, electrochemical methods (Tafel and EIS), and SEM	–	3 vol%	[73]
74	Banana leaves water extract	(S)-2-amino-3–(4-hydroxyphenyl) propanoic acid (AHP), 3–(2-aminoethyl)-1H-indol-5-ol (ATI), and 3,4-dihydroxy-cinnamic acid (DDA)	X70 steel at 1 mol/L hydrochloric acid	Electrochemical methods and theoretical studies (QC and MD)	90%	-	[72]
75	*Centipeda minima*	Ethyl 3–(3, 4–dihydroxyphenyl) acrylate (EDA), 3',4',5,7–tetrahydroxy–3–methoxy flavone (THM), 1,2–bis(3,5–dimethoxyphenyl) ethylene (BDE), and apigenin (AGN)	Carbon steel in 1.0 mol/L HCl	Electrochemical methods and theoretical studies (QC and MD)	96.2%	500 mg/L	[78]
76	*Cinnamoum tamala*	Eugenol, (c) methyl eugenol, and (d) cinnamyl	Low-carbon steel (LCS) in the solution of 0.5 M H_2SO_4	Weight loss, electrochemical methods, and theoretical studies (QC and MD)	96.76%	100 mg/L	[79]
77	*Citrus aurantium*	N.M	Mild steel in 1 M H_2SO_4	Weight loss	89%	10 mL/L	[77]

78	*Nicotiana tabacum*	Nicotine	Mild steel in 2 M H_2SO_4	Weight loss and quantum chemical study	The %IE and corrosion resistance of the mild steel increased with inhibitor concentrations	–	[80]
79	*Diospyros kaki* (persimmon)	N.M	St37 steel corrosion in 0.1 M HCl	Electrochemical methods (Tafel and EIS)	91%	225 ppm	[81]
80	*Garcinia livingstonei*	Tannins, flavonoids, reducing sugars, terpenoids, phenolic compounds, steroid compounds, and glycoside compounds	Mild steel in 1 M HCl	Weight loss, Tafel, EIS, and SEM	95.23%	4.0 g/L	[82]
81	*Paederia foetida*	N.M	Mild steel in 1 M hydrochloric acid	Weight loss, Tafel, EIS, and SEM	73.77%	–	[83]
82	*Bambusa arundinacea*	N.M	Steel-reinforced concrete	–	–	–	[87]
83	Henna	N.M	Acrylic resin coating	Electrochemical methods (Tafel and EIS)	99.9%	0.2 wt/vol%	[88]
84	*Lilium brownii*	N.M	X70 steel in 1 M HCl solution	Electrochemical methods (Tafel and EIS) and theoretical calculations	85%	200 mg/L	[89]
85	*Morus alba pendula*	N.M	Carbon steel in 1 M HCl solution	Electrochemical methods (Tafel and EIS)	93%	0.4 g/L	[90]
86	Hybrid pigment based on zinc acetate–*Cichorium intybus L* leaf extract	Chicoric acid, caffeic acid, flavone	Mild steel in aqueous chloride solutions	Electrochemical methods (Tafel and EIS)	93%	–	[86]

(*Continued*)

TABLE 3.1 (*Continued*)
The Detail of Plant Extracts as a Corrosion Inhibitor

Num.	Plant Extract	Anticorrosion Active Constituents	System	Techniques Used	Highest Protection Efficiency (%)	Inhibitor Concentration	Reference
87	Papaya	DL-ascorbic acid(DLA), 5,7-dimethoxy-chromen-2-one (DCO), 2-amino-4-methylsulfanyl-butyric acid (AMB), 3,4-ditydroxy-benzoic acid(DBA).), 2-hydroxy-succinic acid (HSA), 3–[3-(3,4-dihydroxy-phenyl)-acryloxy]-1,4,5-trihydroxy-cyclohexane-carboxylic acid (DTA)	Cu in 0.5 M H_2SO_4	Electrochemical methods (Tafel and EIS)	95.5%	150 mg/L	[91]
88	*Psidium guajava Linn.*	(S)-2-amino-5-guanidnopentanoic acid (AGPA), 2–(3,4-dihydroxyphenyl)-3,5,7-trihydroxy-4H-chromen-4-one (DDTC),(S)-2-amino-3-phenylpropanoic acid (APPA), and (2R,3S)-2–(3,4-dihydroxyphenyl) chroman-3,5,7-trial (DPCT)	Copper in H_2SO_4 environment	Electrochemical methods (Tafel and EIS)	95.1%	600 mg/L	[92]
89	Zinc acetylacetonate/*Urtica dioica* leaves	Histamine, serotonin, and caffeic acid	Mild steel in 3.5% NaCl	Tafel	–	–	[131]
90	*Ficus hispida*	Stigmasterol	Steel in 1 M HCl	Weight loss, Tafel, EIS, and SEM	90%	250 ppm	[84]
91	*Dolichandra unguis-cati*	–	Mild steel in 1 M HCl	Weight loss, Tafel, EIS, and SEM	93.61%	0.76 g/L	[85]
92	*Xanthium strumarium*	–	Low-carbon steel in 1 M HC	Weight loss	94.82%	10 mL/L	[93]

			Stem				
93	*Cnicus benedictus* weed	Chlorogenic acid and sinapic acid	Mild steel corrosion in 0.5 M HCl	Weight loss, Tafel, EIS, and DFT	92.45%	1,000 ppm	[94]
94	*Corchorus olitorius*	–	Mild steel in 0.5 M H_2SO_4	Weight loss	94.34%	1.16 g/L	[95]
95	Zinc cations/ *Nepeta pogonosperma*, locally called larzeh (LA)	Terpinen-4-ol, α-terpineol, β-pinene, 1,8-cineole, and 4aα-7α-7aα-nepetalactone	Mild steel in the saline solution	Electrochemical methods (Tafel and EIS)	94%	300:700 ppm LA:Zn solution	[96]
96	Pineapple stem extract (bromelain)	–	Low-carbon steel in 1 M HCl	Weight loss, Tafel, and EIS	97.60%	1,000 ppm	[97]
97	Brahmi (*Bacopa monnieri*) and Henna	Bacoside-A and bacoside-B	Low-carbon steel in 0.5 M NaOH	Weight loss and Tafel	80%	0.5–2%	[98]
			Gum				
98	Almond	–	Mild steel in 1 M HCl	Weight loss and electrochemical methods (Tafel and EIS) and theoretical studies (DFT and MC)	96.37%	300 ppm	[100]
99	*Boswellia serrata*	Xylose, arabinose, galactose, and glucuronic acid	Low-carbon steel in 1 M HCl	Weight loss and electrochemical methods (Tafel and EIS)	91.90%	500 ppm	[101]
100	1:1 combination of Arabic and cashew	[6-O-(1-L-maloyl)-alpha-, beta-D-Glcp (veracylglucan A), 6-octadecanoic acid, methylhexadecanoate, 1-pentadecane carboxylic acid, and 9, 12-octadecadienoic acid	Mild steel	Weight loss	83.12%	0.5 g/L	[102]

(*Continued*)

TABLE 3.1 (*Continued*)
The Detail of Plant Extracts as a Corrosion Inhibitor

Num.	Plant Extract	Anticorrosion Active Constituents	System	Techniques Used	Highest Protection Efficiency (%)	Inhibitor Concentration	Reference
101	Sugar beet pectin	–	Mild steel in HCl	Weight loss and electrochemical methods (Tafel and EIS)	93.9%	1 g/L	[44]
102	*Raphia hookeri*	–	Aluminum in HCl	Weight loss	56.30%	0.5 g/L	[103]
103	Polyacrylamide-grafted guar gum	–	Mild steel in 1 M HCl	Electrochemical methods (Tafel and EIS)	90%	1%	[106]
104	Guar gum and ethyl acrylate composite	–	P110 steel corrosion protection in 15% HCl	Weight loss and electrochemical methods (Tafel and EIS)	92.30%	500 mg/L	[107]
105	Guar gum–grafted 2-acrylamido-2-methylpropanesulfonic acid (GG-AMPS)	–	Carbon steel in ASTM standard D1141–98(2013) artificial seawater	Gravimetric (mass loss), theoretical simulations, and electrochemical	92.5%	600 mg/L	[108]
106	Guar gum	–	Carbon steel corrosion in 2 M H3PO4 solution	Weight loss, electrochemical methods (Tafel and EIS), and molecular dynamic (MD)	95.80%	1 g/L	[109]
107	Surfactant addition to guar gum (GG)	–	Mild steel in 0.5 M HCl	Weight loss and electrochemical methods (Tafel and EIS)	97.13%	800:5 ppm	[110]

108	Gum acacia	–	Mild steel in 15% HCl	Weight loss and electrochemical methods (Tafel and EIS)	94.08%	0.3 g/L	[111]
109	Terebinth	a-pinene, limonene, and a-terpineol	Iron in 3% NaCl medium	Electrochemical methods (Tafel and EIS)	86.40%	3,000 ppm	[112]
110	*Ferula assa-foetida* and *Dorema ammoniacum oleo* gum resin exudate	Umbelliferone, ferulic acid, farnesiferol-C, farnesiferol-B, 1-propenyl propyl disulfide, (R)-2-butyl-1-propenyl disulfide, and 1(1-methylthiopropyl) -1-propenyl disulfide in *F. assa-foetida*, and linalool, linalyl acetate, citronellyl acetate, and ammoresinol in *D. ammoniacum*	Mild steel in 2 M HCl	Weight loss and electrochemical methods (Tafel and EIS)	–	–	[113]
111	Xanthan gum and xanthan gum–grafted poly(acrylamide)	–	Mild steel in 15% HCl	Weight loss and electrochemical methods (Tafel and EIS)	92.80%	0.4 g/L	[114]
112	Morocco, Mauritania, and Senegal gum Arabic variants	–	Mild steel in 1 M HCl	Weight loss, electrochemical methods (Tafel and EIS) and DFT	96%, 95%, and 94% for Morocco, Mauritania, and Senegal	1 g/L	[115]
113	*Lupinus arboreus*	Saponins, glycoside, flavonoid, steroid, tannin	Epoxy-based coating	EIS	–	–	[116]
114	Locust bean	–	Q235 steel in 0.5 M H_2SO_4	Electrochemical methods (Tafel and EIS)	89.80%	5 mM	[117]

(Continued)

TABLE 3.1 (*Continued*)
The Detail of Plant Extracts as a Corrosion Inhibitor

Num.	Plant Extract	Anticorrosion Active Constituents	System	Techniques Used	Highest Protection Efficiency (%)	Inhibitor Concentration	Reference
			Flower				
115	Areca	–	Al in 0.5 M HCl	Weight loss and electrochemical methods (Tafel and EIS)	94%	18 g/L	[118]
116	Bee pollen	Gallic acid, vanillic acid, rutin, D-gluconic acid, linoleic acid ethyl ester, stearic acid, behenic acid, oleanolic acid, propanoic acid, myristic acid, benzenepropanoic acid, 9,12,15-octadecatrienoic acid, luteolin, kaempferol, and palmitic acid	Pure copper in 1 M HCl	Weight loss and electrochemical methods (Tafel and EIS)	94.50%	7 g/L	[119]
117	*Borassus flabellifer* coir dust thus extracted by water (BFDWE) and methanol (BFDME)	–	Al in 1 M H_2SO_4	Weight loss and electrochemical methods (Tafel and EIS)	66.8% and 52% for the BFDME and BFDWE, respectively	0.4 g/L	[120]
118	Dardagan fruit	–	Mild steel in 1 M HCl	Electrochemical methods (Tafel and EIS)	97%	3,000 ppm	[121]
119	*Cistus monspeliensis L* leaves, flowers, and stems extract	–	Mild steel in 1 M HCl	Weight loss and electrochemical methods (Tafel and EIS)	88%, 81%, and 70%	0.2 g/L CMAE leaves, flowers, and stems, respectively	[122]

120	Primrose flower and zinc cations	Cinnamic acid, protocatechuic acid, and catechin	Mild steel in 3.5 wt% NaCl	Electrochemical methods (Tafel and EIS) and molecular simulations	95.30%	Combination of 300 ppm PPE + 700 ppm Zn	[123]
121	*Primula vulgaris* flower aqueous extract	Catechin, cinnamic acid, protocatechuic acid, and rutin	Mild steel in 1 M HCl	EIS, polarization, and theoretical studies (DFT, MD, and MC)	95.51%	1,000 ppm	[124]
122	*Stachys byzantina* and zinc (II) cation	–	Carbon steel in 3.5% NaCl solution	Electrochemical methods (Tafel and EIS)	92%	700 ppm SB and 300 ppm of ZN	[125]
123	Wine industry residues	–	API P110 steel in 1 M HCl	Weight loss and electrochemical methods (Tafel and EIS)	83.81%	4.0 g/L	[126]
124	*Tagetes erecta* (marigold flower)	Lutein	Mild steel in 0.5 M H_2SO_4	Weight loss and electrochemical methods (Tafel and EIS)	98.07%	1.0 g/L	[127]
125	*Thevetia peruviana*	Cannogenin, connonigenol, digitoxigenin, and thevetin B	Mild steel in 1 M HCl	EIS, polarization, and theoretical studies (DFT and MD)	91.24%	200 mg/L	[128]
126	*Tithonia diversifolia*	–	Mild steel in 1 M HCl	Weight loss and electrochemical methods (Tafel and EIS)	91.24%	200 mg/L	[129]
127	*Ziziphus spina-christi*	–	Carbon steel in acidic conditions	Weight loss and electrochemical methods (Tafel and EIS)	91.35%	810 ppm	[130]

3.18 ELECTROCHEMICAL CORROSION INVESTIGATION METHODS

Corrosion and corrosion inhibitors function can be investigated by many methods. The most common ones are classical (e.g., weight loss) and electrochemical methods, including impedance spectroscopy (EIS), anodic polarization, linear polarization, cyclic polarization, Tafel polarization, field effect method, quantum studies, and electron imaging. Here we discuss impedance spectroscopy and Tafel polarization methods.

3.19 TAFEL POLARIZATION INVESTIGATION

Corrosion is a spontaneous process and a set of oxidation and reduction reactions which are governed by electrochemical laws, so electrochemical methods for evaluating these processes and checking the performance of inhibitors are very accurate. One of the most important methods for studying corrosion and determining the mechanism of this process by measuring very low corrosion rates is the polarization method. The deviation of the electrode potential or cell potential from the reversible value (i.e., the equilibrium value) with the passage of Faraday current is called polarization. The higher the deviation, the higher the polarization. Polarization can exist in two ways:

- *Activation polarization.* This type of polarization is created by the inhibitory factors that exist in the reaction environment.
- *Concentration polarization.* Due to the passage of electric current and the change in the concentration of ions near the surface of the metal, the speed of the electrochemical reaction is reduced and the speed of the reaction is controlled by the speed of penetration of the ions to the surface of the metal. At this time, stirring the solution increases the rate of penetration and decreases the polarization and increases the rate of corrosion.

One of the first points raised in relation to corrosion is the corrosion potential, and current, which must be tested and reported in each case. The potential at which the cathode and anode reach equilibrium with each other is called the corrosion potential, which is represented by *Ecorr.* The amount of current at this point is also called corrosion current and is indicated by I_{corr}. Therefore, it is necessary to address how to achieve the mentioned items through electrochemistry and the relevant device. The most important method of examining I_{corr} experimentally is the extraction of Tafel charts, first used by Trude and Wagner. To draw the Tafel curve, the open circuit corrosion potential (E_{corr}) should be determined first, then the potential should start from a negative potential compared to the corrosion potential and continue until a positive potential compared to the corrosion potential. In the first stage, the cathodic Tafel curve, and in the next step the anodic curve, is drawn. Finally, the potential curve is drawn according to the logarithm of the current density. To determine the corrosion current, the linear part of the Tafel curve (anodic or cathodic) must be continued until it intersects the horizontal line crossing E_{corr}. The reason for the use of the linear part of the Tafel curve is that only at low overpotentials are the reactions under kinetic control, and at large overpotentials, the Tafel curve becomes a smooth line, where the reactions are under mass transfer control. Anode metal surface, I_{corr} can be used to calculate the corrosion current density, corrosion intensity, and corrosion penetration rate. The corrosion rate value is obtained from the intersection of the linear extension of the Tafel curve and the horizontal line passing through the corrosion potential. The Tafel constants are also obtained from the slope. Anodic and cathodic branches are obtained (Figure 3.9) [132].

Finally, by comparing the values of the parameters obtained from the investigation of metal corrosion in the control solution with the corresponding values of the solutions containing the inhibitor, the function of the inhibitors on the metal surface can be checked.

3.19.1 Calculation of Inhibition Percentage

From the comparison of the corrosion current density resulting from the corrosion of the metal in the control solution compared to the solution containing the inhibitor, the inhibition percentage of the inhibitor is calculated using equation 1.

$$IE_P(\%) = (1 - \left(\frac{I_{corr}}{I^o_{corr}}\right)) \times 100$$

In this regard, $I^o corr$ and *Icorr* are the corrosion current densities (in A cm[2]) in the absence and presence of the inhibitor, respectively.

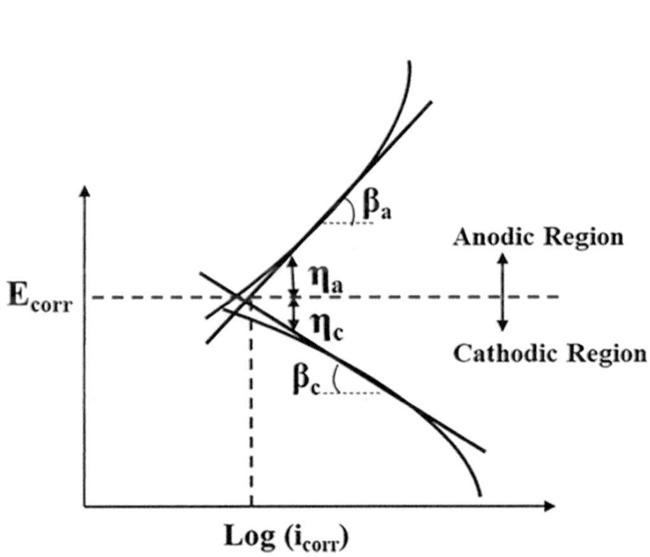

FIGURE 3.9 Anodic and cathodic branches, potential (E_{corr}), corrosion current (I_{corr}), and linear extrapolation in Tafel diagram.

3.19.2 Determining Inhibitor Performance on the Mechanism of Cathodic and Anodic Reactions

The changes in the slope of the Tafel lines in the presence of the inhibitor compared to the control medium show that the inhibitor has changed the corrosion mechanism, on which of the cathodic and anodic reactions or both of them.

3.19.3 Determining the Type of Inhibitor (Cationic, Anionic, or Mixed)

Anodic inhibitors are substances that reduce the current of the anodic branch compared to the control, while the current of the cathodic branch is almost the same for the control solution and the solutions containing the inhibitor. In this case, the corrosion potential of the solution containing the inhibitor is shifted to more positive values compared to the potential of the control solution (Figure 3.10a). Cathodic inhibitors are substances that reduce the current of the cathodic branch compared to the control solution, but there is no significant change in the current of the anodic branch. In this case, the corrosion potential shifts significantly toward negative potentials. These inhibitors cannot remove the metal surface like the oxidizing inhibitors (Figure 3.10b). Mixed inhibitors are substances that reduce the flow of both cathodic and anodic branches compared to the control solution and do not cause a significant change in the corrosion potential. The performance of these inhibitors is much better than that of anodic and cathodic inhibitors. These types of materials mostly contain nitrogen, sulfur, or both elements, and by being absorbed on the surface of the metal, they cause a delay in the cathodic phenomena and slow down the anodic phenomena (Figure 3.10-C).

3.20 ELECTROCHEMICAL IMPEDANCE SPECTROSCOPY INVESTIGATION

In alternating voltage systems where the current and potential change sinusoidally and have a phase difference of $\pi/2$ with respect to each other, the alternating resistance caused by this system is called impedance. The impedance method is used to

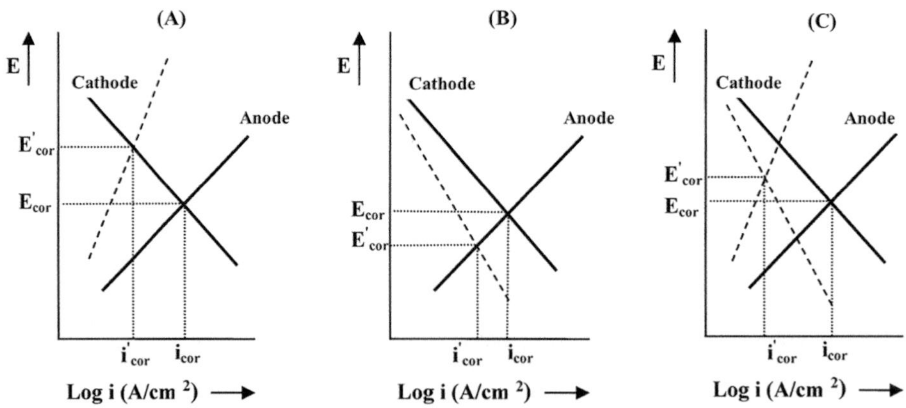

FIGURE 3.10 Cathodic, anodic, and mixed inhibitory effect.

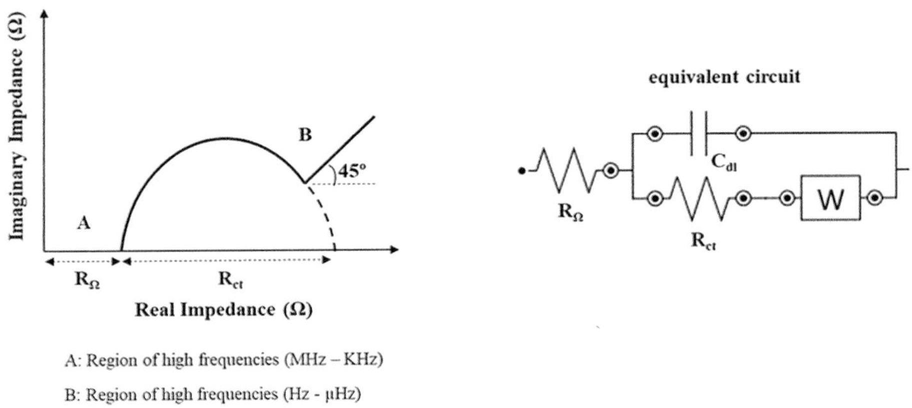

FIGURE 3.11 An example of an EIS spectrum and its equivalent circuit.

check the kinetics of electron transfer reactions; therefore, according to Warberg's studies, the impedance should be checked at high frequencies, where the system is under kinetic control. At low frequencies, the electroactive species has a lot of opportunity to be oxidized and reduced, and electrolysis occurs on the surface of the cathode electrode. In this case, the thickness of the penetration layer increases and the system will be under penetration control. So to check the impedance, move from high to low frequencies so that the kinetics of electron transfer can be checked. Also, in the investigation of kinetics with the help of impedance, a very small overpotential should be applied to the system compared to the inherent potential (Nernst relation potential); otherwise, the system will have a twist in the current potential sinusoidal response. Electron transfer reactions are investigated by using changes in Rct (resistance of charge transfer or polarization). During the application of the potential, an electric double layer is formed at the solution–metal interface, which is represented by *Cdl*. Also, *RΩ* (uncompensated resistance) is formed due to the resistance of the solution between the working electrode and the reference electrode. The existence of these two components makes the calculation work more difficult. Therefore, the mixed-plane impedance analysis method will be used. Impedance graphs are drawn as real impedance (Z′) based on imaginary impedance (Z″). Then a comparison is made between the considered electrochemical cell and an equivalent electric circuit, including capacitors and resistors [132]. Figure 3.11 shows the simplest type of equivalent circuit along with an EIS spectrum.

REFERENCES

1. Tang, Z., A review of corrosion inhibitors for rust preventative fluids. *Current Opinion in Solid State and Materials Science*, 2019. **23**(4): p. 100759.
2. Raja, P.B. and M.G. Sethuraman, Natural products as corrosion inhibitor for metals in corrosive media – a review. *Materials Letters*, 2008. **62**(1): p. 113–116.
3. Dehghani, A., et al., A combined experimental and theoretical study of green corrosion inhibition of mild steel in HCl solution by aqueous Citrullus lanatus fruit (CLF) extract. *Journal of Molecular Liquids*, 2019. **279**: p. 603–624.

4. Cherrad, S., et al., Cupressus arizonica fruit essential oil: A novel green inhibitor for acid corrosion of carbon steel. *Arabian Journal of Chemistry*, 2022. **15**(6).
5. Rani, A.T.J., et al., The influence of aqueous and alcoholic extracts of Garcinia cambogia fruit rind in the management of mild steel corrosion in hydrochloric acid: Theoretical and electroanalytical studies. *Journal of Molecular Liquids*, 2022. **346**.
6. Thomas, A., J.R. At, and A. Joseph, Extended protection of mild steel in molar HCl using the Garcinia Indica fruit rind extract (GIW) and iodide ions; electrochemical, thermodynamic and kinetic studies. *Journal of the Indian Chemical Society*, 2021. **98**(10).
7. Liao, L.L., et al., Corrosion protection for mild steel by extract from the waste of lychee fruit in HCl solution: Experimental and theoretical studies. *Journal of Colloid and Interface Science*, 2018. **520**: p. 41–49.
8. Anupama, K.K., et al., Adsorption and electrochemical studies of Pimenta dioica leaf extracts as corrosion inhibitor for mild steel in hydrochloric acid. *Materials Chemistry and Physics*, 2015. **167**: p. 28–41.
9. Olasehinde, E.F., et al., Corrosion protection of mild steel in acidic medium by titanium-based nanocomposite of Chromolaena odorata leaf extract. *Materials Chemistry and Physics*, 2022. **281**.
10. Nathiya, R.S. and V. Raj, Evaluation of Dryopteris cochleata leaf extracts as green inhibitor for corrosion of aluminium in 1 M H 2 SO 4. *Egyptian Journal of Petroleum*, 2017. **26**(2): p. 313–323.
11. Dehghani, A., G. Bahlakeh, and B. Ramezanzadeh, Green Eucalyptus leaf extract: A potent source of bio-active corrosion inhibitors for mild steel. *Bioelectrochemistry*, 2019. **130**: p. 107339.
12. Qiang, Y., et al., Evaluation of Ginkgo leaf extract as an eco-friendly corrosion inhibitor of X70 steel in HCl solution. *Corrosion Science*, 2018. **133**: p. 6–16.
13. Mobin, M., M. Basik, and Y. El Aoufir, Corrosion mitigation of mild steel in acidic medium using Lagerstroemia speciosa leaf extract: A combined experimental and theoretical approach. *Journal of Molecular Liquids*, 2019. **286**.
14. Muthukrishnan, P., B. Jeyaprabha, and P. Prakash, Adsorption and corrosion inhibiting behavior of Lannea coromandelica leaf extract on mild steel corrosion. *Arabian Journal of Chemistry*, 2017. **10**: p. S2343–S2354.
15. Rahal, C., et al., Olive leaf extract as natural corrosion inhibitor for pure copper in 0.5 M NaCl solution: A study by voltammetry around OCP. *Journal of Electroanalytical Chemistry*, 2016. **769**: p. 53–61.
16. Zhou, Y., et al., Combining electrochemical, surface topography analysis, and theoretical calculation methods to insight into the anti-corrosion property of Syzygium samarangense leaf extract. *Journal of Industrial and Engineering Chemistry*, 2021. **102**: p. 302–311.
17. Fernandes, F.D., L.M. Ferreira, and M.L.C.P. da Silva, Evaluation of the corrosion inhibitory effect of the ecofriendily additive of Terminalia Catappa leaf extract added to soybean oil biodiesel in contact with zinc and carbon steel 1020. *Journal of Cleaner Production*, 2021. **321**.
18. Silva, M.V.L.D., E.D.B. Policarpi, and A. Spinelli, Syzygium cumini leaf extract as an eco-friendly corrosion inhibitor for carbon steel in acidic medium. *Journal of the Taiwan Institute of Chemical Engineers*, 2021. **129**: p. 342–349.
19. Harb, M.B., et al., Olive leaf extract as a green corrosion inhibitor of reinforced concrete contaminated with seawater. *Arabian Journal of Chemistry*, 2020. **13**(3): p. 4846–4856.
20. Soltani, N. and M. Khayatkashani, Evaluating performance of malva sylvestris leaf extract for protection of mild steel against corrosion in acidic solution. *Asian Journal of Green Chemistry*, 2021. **5**: p. 39–57.
21. El Hamdani, N., et al., Alkaloids extract of Retama monosperma (L.) Boiss. seeds used as novel eco-friendly inhibitor for carbon steel corrosion in 1 M HCl solution: Electrochemical and surface studies. *Applied Surface Science*, 2015. **357**: p. 1294–1305.

22. Gallia, M.C., et al., Pehuen (Araucaria araucana) seed residues are a valuable source of natural antioxidants with nutraceutical, chemoprotective and metal corrosion-inhibiting properties. *Bioorganic Chemistry*, 2020. **104**: p. 104175.
23. Damej, M., et al., An environmentally friendly formulation based on Cannabis sativa L. seed oil for corrosion inhibition of E24 steel in HCl medium: Experimental and theoretical study. *Colloids and Surfaces A: Physicochemical and Engineering Aspects*, 2022. **643**.
24. Abbout, S., et al., Ceratonia Siliqua L seeds extract as eco-friendly corrosion inhibitor for carbon steel in 1 M HCl: Characterization, electrochemical, surface analysis, and theoretical studies. *Journal of Molecular Structure*, 2021. **1240**.
25. Dehghani, A., et al., Detailed macro-/micro-scale exploration of the excellent active corrosion inhibition of a novel environmentally friendly green inhibitor for carbon steel in acidic environments. *Journal of the Taiwan Institute of Chemical Engineers*, 2019. **100**: p. 239–261.
26. Majd, M.T., et al., Production of an environmentally stable anti-corrosion film based on Esfand seed extract molecules-metal cations: Integrated experimental and computer modeling approaches. *Journal of Hazardous Materials*, 2020. **382**: p. 121029.
27. Haldhar, R., et al., Evaluation of Gloriosa superba seeds extract as corrosion inhibition for low carbon steel in sulfuric acidic medium: A combined experimental and computational studies. *Journal of Molecular Liquids*, 2021. **323**.
28. Njoku, D.I., et al., Corrosion protection of Q235 steel in acidic-chloride media using seed extracts of Piper guineense. *Journal of Molecular Liquids*, 2021. **330**.
29. Wang, D.-Y., et al., Fabricating of grape seed proanthocyanidins loaded Zein-NaCas composite nanoparticles to exert effective inhibition of Q235 steel corrosion in seawater. *Journal of Molecular Liquids*, 2022. **348**.
30. Mostafatabar, A.H., et al., A comprehensive electronic-scale DFT modeling, atomic-level MC/MD simulation, and electrochemical/surface exploration of active nature-inspired phytochemicals based on Heracleum persicum seeds phytoextract for effective retardation of the acidic-induced corrosion of mild steel. *Journal of Molecular Liquids*, 2021. **331**.
31. Pal, S., et al., Lemon seeds as green coating material for mitigation of mild steel corrosion in acid media: Molecular dynamics simulations, quantum chemical calculations and electrochemical studies. *Journal of Molecular Liquids*, 2020. **316**.
32. Bahlakeh, G., et al., Highly effective mild steel corrosion inhibition in 1 M HCl solution by novel green aqueous Mustard seed extract: Experimental, electronic-scale DFT and atomic-scale MC/MD explorations. *Journal of Molecular Liquids*, 2019. **293**.
33. Fernine, Y., et al., Anti-corrosion performance of Ocimum basilicum seed extract as environmental friendly inhibitors for mild steel in HCl solution: Evaluations of electrochemical, EDX, DFT and Monte Carlo. *Journal of Molecular Liquids*, 2022. **355**.
34. Bahlakeh, G., et al., Novel cost-effective and high-performance green inhibitor based on aqueous Peganum harmala seed extract for mild steel corrosion in HCl solution: Detailed experimental and electronic/atomic level computational explorations. *Journal of Molecular Liquids*, 2019. **283**: p. 174–195.
35. Bhardwaj, N., et al., Phyllanthus emblica seed extract as corrosion inhibitor for stainless steel used in petroleum industry (SS-410) in acidic medium. *Chemical Physics Impact*, 2021. **3**.
36. Shahmoradi, A.R., et al., Molecular-MD/atomic-DFT theoretical and experimental studies on the quince seed extract corrosion inhibition performance on the acidic-solution attack of mild-steel. *Journal of Molecular Liquids*, 2022. **346**.
37. Mobtaker, H., et al., The inhibition performance of quinoa seed on corrosion behavior of carbon steel in the HCl solution; theoretical and experimental evaluations. *Journal of Molecular Liquids*, 2021. **335**.

74. Sin, H.L.Y., et al., Aquilaria subintergra leaves extracts as sustainable mild steel corrosion inhibitors in HCl. *Measurement*, 2017. **109**: p. 334–345.
75. Arifa Farzana, B., A. Mushira Banu, and K. Riaz Ahamed, Andrographis echioides leaves extract as an eco-friendly corrosion inhibitor for mild steel in acid medium. *Materials Today: Proceedings*, 2021. **47**: p. 2080–2090.
76. Guruprasad, A.M. and H.P. Sachin, Novel cost-effective aqueous Amorphophallus paeoniifolius leaves extract as a green corrosion inhibitor for mild steel corrosion in hydrochloric acid medium: A detailed experimental and surface characterization studies. *Chemical Data Collections*, 2021. **34**.
77. Hassan, K.H., A.A. Khadom, and N.H. Kurshed, Citrus aurantium leaves extracts as a sustainable corrosion inhibitor of mild steel in sulfuric acid. *South African Journal of Chemical Engineering*, 2016. **22**: p. 1–5.
78. Wang, Q., et al., Insight into anti-corrosion behavior of Centipeda minima leaves extract as high-efficiency and eco-friendly inhibitor. *Colloids and Surfaces A: Physicochemical and Engineering Aspects*, 2022. **640**.
79. Prasad, D., et al., Cinnamoum tamala leaves extract highly efficient corrosion bio-inhibitor for low carbon steel: Applying computational and experimental studies. *Journal of Molecular Liquids*, 2022. **347**.
80. Bhawsar, J., P.K. Jain, and P. Jain, Experimental and computational studies of Nicotiana tabacum leaves extract as green corrosion inhibitor for mild steel in acidic medium. *Alexandria Engineering Journal*, 2015. **54**(3): p. 769–775.
81. Gerengi, H., et al., Evaluation of the inhibitive effect of Diospyros kaki (Persimmon) leaves extract on St37 steel corrosion in acid medium. *Sustainable Chemistry and Pharmacy*, 2016. **4**: p. 57–66.
82. Rathod, M.R., S.K. Rajappa, and A.A. Kittur, Garcinia livingstonei leaves extract influenced as a mild steel efficient green corrosion inhibitor in 1 M HCl solution. *Materials Today: Proceedings*, 2022. **54**: p. 786–796.
83. Hossain, N., et al., Paederia Foetida leaves extract as a green corrosion inhibitor for mild steel in hydrochloric acid solution. *Current Research in Green and Sustainable Chemistry*, 2021. **4**.
84. Muthukrishnan, P., et al., Stigmasterol extracted from Ficus hispida leaves as a green inhibitor for the mild steel corrosion in 1 M HCl solution. *Arabian Journal of Chemistry*, 2019. **12**(8): p. 3345–3356.
85. Rathod, M.R., et al., Investigation of Dolichandra unguis-cati leaves extract as a corrosion inhibitor for mild steel in acid medium. *Current Research in Green and Sustainable Chemistry*, 2021. **4**.
86. Sanaei, Z., T. Shahrabi, and B. Ramezanzadeh, Synthesis and characterization of an effective green corrosion inhibitive hybrid pigment based on zinc acetate-Cichorium intybus L leaves extract (ZnA-CIL.L): Electrochemical investigations on the synergistic corrosion inhibition of mild steel in aqueous chloride solutions. *Dyes and Pigments*, 2017. **139**: p. 218–232.
87. Asipita, S.A., et al., Green Bambusa Arundinacea leaves extract as a sustainable corrosion inhibitor in steel reinforced concrete. *Journal of Cleaner Production*, 2014. **67**: p. 139–146.
88. Zulkifli, F., et al., Henna leaves extract as a corrosion inhibitor in acrylic resin coating. *Progress in Organic Coatings*, 2017. **105**: p. 310–319.
89. Zuo, X., et al., Research of Lilium brownii leaves extract as a commendable and green inhibitor for X70 steel corrosion in hydrochloric acid. *Journal of Molecular Liquids*, 2021. **321**.
90. Jokar, M., T.S. Farahani, and B. Ramezanzadeh, Electrochemical and surface characterizations of morus alba pendula leaves extract (MAPLE) as a green corrosion inhibitor for steel in 1M HCl. *Journal of the Taiwan Institute of Chemical Engineers*, 2016. **63**: p. 436–452.

91. Tan, B., et al., Papaya leaves extract as a novel eco-friendly corrosion inhibitor for Cu in H(2)SO(4) medium. *Journal of Colloid and Interface Science*, 2021. **582**(Pt B): p. 918–931.
92. Wu, Y., L. Guo, and Y. She, Insight on the corrosion inhibition performance of psidium guajava linn leaves extract. *Journal of Molecular Liquids*, 2022. **346**.
93. Khadom, A.A., A.N. Abd, and N.A. Ahmed, Xanthium strumarium leaves extracts as a friendly corrosion inhibitor of low carbon steel in hydrochloric acid: Kinetics and mathematical studies. *South African Journal of Chemical Engineering*, 2018. **25**: p. 13–21.
94. Thakur, A., et al., Computational and experimental studies on the corrosion inhibition performance of an aerial extract of Cnicus Benedictus weed on the acidic corrosion of mild steel. *Process Safety and Environmental Protection*, 2022. **161**: p. 801–818.
95. Oyewole, O., T.A. Oshin, and B.O. Atotuoma, Corchorus olitorius stem as corrosion inhibitor on mild steel in sulphuric acid. *Heliyon*, 2021. **7**(4): p. e06840.
96. Haddadi, S.A., et al., Detailed experimental investigation of the highly active corrosion inhibitive green molecules based on zinc cations/Nepeta Pogonosperma extract and toward the corrosion mitigation of mild steel in the saline solution. *Colloids and Surfaces A: Physicochemical and Engineering Aspects*, 2022. **642**.
97. Mobin, M., M. Basik, and J. Aslam, Pineapple stem extract (Bromelain) as an environmental friendly novel corrosion inhibitor for low carbon steel in 1 M HCl. *Measurement*, 2019. **134**: p. 595–605.
98. Al Hasan, N.H.J., et al., The dual effect of stem extract of Brahmi(Bacopamonnieri) and Henna as a green corrosion inhibitor for low carbon steel in 0.5 M NaOH solution. *Case Studies in Construction Materials*, 2019. **11**.
99. Shahini, M.H., B. Ramezanzadeh, and H.E. Mohammadloo, Recent advances in biopolymers/carbohydrate polymers as effective corrosion inhibitive macro-molecules: A review study from experimental and theoretical views. *Journal of Molecular Liquids*, 2021. **325**.
100. Mobin, M., et al., Experimental and theoretical assessment of almond gum as an economically and environmentally viable corrosion inhibitor for mild steel in 1 M HCl. *Sustainable Chemistry and Pharmacy*, 2020. **18**.
101. Mobin, M., M. Basik, and J. Aslam, Boswellia serrata gum as highly efficient and sustainable corrosion inhibitor for low carbon steel in 1 M HCl solution: Experimental and DFT studies. *Journal of Molecular Liquids*, 2018. **263**: p. 174–186.
102. Arthur, D.E., Computational and experimental study on corrosion inhibition potential of the synergistic 1:1 combination of Arabic and cashew gums on mild steel. *Petroleum Research*, 2020. **5**(2): p. 170–180.
103. Umoren, S.A., et al., The Inhibition of aluminium corrosion in hydrochloric acid solution by exudate gum from Raphia hookeri. *Desalination*, 2009. **247**(1–3): p. 561–572.
104. Mudgil, D., S. Barak, and B.S. Khatkar, Guar gum: Processing, properties and food applications – a review. *Journal of Food Science and Technology*, 2014. **51**: p. 409–418.
105. Mudgil, D., S. Barak, and B. Khatkar, X-ray diffraction, IR spectroscopy and thermal characterization of partially hydrolyzed guar gum. *International Journal of Biological Macromolecules*, 2012. **50**(4): p. 1035–1039.
106. Roy, P., et al., Corrosion inhibition of mild steel in acidic medium by polyacrylamide grafted Guar gum with various grafting percentage: Effect of intramolecular synergism. *Corrosion Science*, 2014. **88**: p. 246–253.
107. Singh, A., et al., Chemically modified guar gum and ethyl acrylate composite as a new corrosion inhibitor for reduction in hydrogen evolution and tubular steel corrosion protection in acidic environment. *International Journal of Hydrogen Energy*, 2021. **46**(14): p. 9452–9465.
108. Singh, A., et al., Corrosion inhibition using guar gum grafted 2-acrylamido-2-methylpropanesulfonic acid (GG-AMPS) in tubular steel joints. *Construction and Building Materials*, 2020. **258**.

109. Messali, M., et al., Guar gum as efficient non-toxic inhibitor of carbon steel corrosion in phosphoric acid medium: Electrochemical, surface, DFT and MD simulations studies. *Journal of Molecular Structure*, 2017. **1145**: p. 43–54.
110. Shamsheera, K.O., et al., Effect of surfactant addition to Guar gum and protection of mild steel in hydrochloric acid at high temperatures: Experimental and theoretical studies. *Journal of Molecular Liquids*, 2021. **331**.
111. Biswas, A., et al., Grafting effect of gum acacia on mild steel corrosion in acidic medium: Gravimetric and electrochemical study. *Journal of Molecular Liquids*, 2018. **251**: p. 470–479.
112. Barbouchi, M., et al., Theoretical modeling and experimental studies of Terebinth extracts as green corrosion inhibitor for iron in 3% NaCl medium. *Journal of King Saud University – Science*, 2020. **32**(7): p. 2995–3004.
113. Behpour, M., et al., The effect of two oleo-gum resin exudate from Ferula assa-foetida and Dorema ammoniacum on mild steel corrosion in acidic media. *Corrosion Science*, 2011. **53**(8): p. 2489–2501.
114. Biswas, A., S. Pal, and G. Udayabhanu, Experimental and theoretical studies of xanthan gum and its graft co-polymer as corrosion inhibitor for mild steel in 15% HCl. *Applied Surface Science*, 2015. **353**: p. 173–183.
115. El Azzouzi, M., et al., Moroccan, Mauritania, and senegalese gum Arabic variants as green corrosion inhibitors for mild steel in HCl: Weight loss, electrochemical, AFM and XPS studies. *Journal of Molecular Liquids*, 2022. **347**.
116. Arukalam, I.O., I.O. Madu, and E.Y. Ishidi, High performance characteristics of Lupinus arboreus gum extract as self-healing and corrosion inhibition agent in epoxy-based coating. *Progress in Organic Coatings*, 2021. **151**.
117. Guo, L., et al., Locust bean gum as a green and novel corrosion inhibitor for Q235 steel in 0.5 M H2SO4 medium. *Journal of Molecular Liquids*, 2020. **310**.
118. Raghavendra, N. and J. Ishwara Bhat, Inhibition of Al corrosion in 0.5 M HCl solution by Areca flower extract. *Journal of King Saud University – Engineering Sciences*, 2019. **31**(3): p. 202–208.
119. Ahmed, R.K. and S. Zhang, Bee pollen extract as an eco-friendly corrosion inhibitor for pure copper in hydrochloric acid. *Journal of Molecular Liquids*, 2020. **316**.
120. Nathiya, R.S., et al., Evaluation of extracts of Borassus flabellifer dust as green inhibitors for aluminium corrosion in acidic media. *Materials Science in Semiconductor Processing*, 2019. **104**.
121. Sedik, A., et al., Dardagan fruit extract as eco-friendly corrosion inhibitor for mild steel in 1 M HCl: Electrochemical and surface morphological studies. *Journal of the Taiwan Institute of Chemical Engineers*, 2020. **107**: p. 189–200.
122. Mechbal, N., et al., Anticorrosive and antioxidant effect of the aqueous extract of the leaves, flowers, and stems of Cistus monspeliensis L: Experimental and computational study. *Journal of Molecular Liquids*, 2021. **331**.
123. Majd, M.T., et al., A detailed investigation of the chloride-induced corrosion of mild steel in the presence of combined green organic molecules of Primrose flower and zinc cations. *Journal of Molecular Liquids*, 2020. **297**.
124. Majd, M.T., et al., Green method of carbon steel effective corrosion mitigation in 1 M HCl medium protected by Primula vulgaris flower aqueous extract via experimental, atomic-level MC/MD simulation and electronic-level DFT theoretical elucidation. *Journal of Molecular Liquids*, 2019. **284**: p. 658–674.
125. Molaeipour, P., M. Ramezanzadeh, and B. Ramezanzadeh, Stachys byzantina extract: A green biocompatible molecules source for graphene skeletons generation on the carbon steel for superior corrosion mitigation. *Bioelectrochemistry*, 2022. **143**: p. 107970.
126. Furtado, L.B., et al., Storage time evaluation of a residue from wine industry as a microencapsulated corrosion inhibitor for 1 M HCl. *Materials Chemistry and Physics*, 2020. **256**.

127. Mourya, P., S. Banerjee, and M.M. Singh, Corrosion inhibition of mild steel in acidic solution by Tagetes erecta (Marigold flower) extract as a green inhibitor. *Corrosion Science*, 2014. **85**: p. 352–363.
128. Haque, J., et al., Corrosion inhibition of mild steel in 1M HCl using environmentally benign Thevetia peruviana flower extracts. *Sustainable Chemistry and Pharmacy*, 2021. **19**.
129. Divya, P., et al., Tithonia Diversifolia flower extract as green corrosion inhibitor for mild steel in acid medium. *Materials Today: Proceedings*, 2019. **18**: p. 1581–1591.
130. Rajamohan, N., et al., Eco-friendly biomass from Ziziphus spina-christi for protection of carbon steel in acidic conditions – parameter effects and corrosion mechanism studies. *Chemosphere*, 2022. **291**(Pt 3): p. 132756.
131. Abrishami, S., R. Naderi, and B. Ramezanzadeh, Fabrication and characterization of zinc acetylacetonate/Urtica Dioica leaves extract complex as an effective organic/inorganic hybrid corrosion inhibitive pigment for mild steel protection in chloride solution. *Applied Surface Science*, 2018. **457**: p. 487–496.
132. Wenzel, T.J., Douglas A. Skoog, Donald M. West, F. James Holler, and Stanley R. Crouch: Fundamentals of analytical chemistry. *Analytical and Bioanalytical Chemistry*, 2013. **25**(405): p. 7903–7904.

4 Measurement of Natural Compounds Using Voltammetric Methods

Melika Sadat Dadgar[1],
Mohammad Hadi Karimi Harandi[2],
Maryam Khayatkashani[3], *and Nasrin Soltani*[4]

[1] Department of Analytical Chemistry, Faculty of Chemistry, University of Kashan, Kashan, Iran.

[2] Department of Chemistry, Imam Hossein University, Tehran, Iran.

[3] Advanced Medical Pharma (AMP-Biotec), Biopharmaceutical Innovation Centre, Via Cortenocera, San Salvatore Telesino, (BN), Italy.

[4] Department of Chemistry, Payam Noor University, Isfahan, Iran.

4.1 INTRODUCTION

Voltammetry is one of the electrochemical analysis methods in which the samples are measured by measuring the current as a function of potential.

In the last five decades, voltammetry methods have become a popular tool for studying electrochemical reactions, electrochemical free radicals, enzymatic catalysis, environmental monitoring, industrial quality control, and the determination of important minor biological and clinical compounds. Voltammetry is one of the most sensitive analytical methods for measurement; therefore, in addition to qualitative analysis, it is also used in quantitative analysis.

The voltammetry method measures the intensity of the current resulting from an electrochemical reaction on the surface of the working electrode parallel to the gradual change of potential applied to it. In the voltammetry method, due to the presence of current during measurement, in case of inhibiting current from passing through the reference electrode, three-electrode cells should be utilized.

The fixed potential of this electrode will change by reason of the current passing through the reference electrode, and the potential of working electrode cannot be estimated accurately and correctly.

The word *voltammetry* consists of two words: *volt* and *amperometry*.

DOI: 10.1201/9781003229032-4

Accordingly, *voltammetry measurement* consists of electrolysis and drawing of the current potential curve between a working and a reference electrode. Mass transfer in voltammetry usually happens in the following three ways:

1. *Diffusion.* The spontaneous transfer of matter under the influence of chemical potential difference or gradient; mass transfers from areas with higher concentration to areas with lower concentration. *Concentration* means intending to reduce concentration difference.
2. *Convection.* The transfer of the matter to the surface of the electrode by a physical movement, such as stirring the solution or rotating the working electrode.
3. *Migration.* The transfer of matter due to the existence of electrostatic forces between the electrode and the solution.

Distinct advantages of the voltammetric method over other available methods include its high sensitivity, low detection limit, and the possibility of simultaneous measurement, so voltammetry is one of the most sensitive methods for quantitative measurement, so that it can measure very low concentrations of parts per million (ppm), and even parts in billion (ppb).

4.2 NEW MODIFIERS USED IN WORKING ELECTRODES

Electrochemical techniques consist of differential pulse voltammetry (DPV), cyclic voltammetry (CV), and square wave voltammetry [1]. The electrical behaviors of a working electrode surface can prevent the measurement rate of the electrochemical reaction of the materials due to its homogenous characteristics and decreased reaction rate and also hinder the efficient transport of electrons between the sample and the electrode [2]. Modified electrodes have a higher selectivity for research demands; therefore, the electrode's performance and sensitivity should be selectively adjusted in comparison to bare electrodes. Working electrodes are modified by fixing materials and substances at the interface with the help of covalent bonding, adsorption, coating methods, electrochemical techniques, etc. Carbon paste–modified electrode is a commonly used one made by applying modifier with the electrode material. The most common modified electrodes are as follows [3].

4.2.1 Carbon Nanomaterial–Modified Electrodes

Nanomaterials have been widely used in modified electrodes by electrochemists. Nanomaterials ranging from 0.1 to 100 nm, or even consisting of fundamental units [4], exhibit special physical and chemical features as a result of quantum size, tiny size, surface impacts, macroscopic quantum tunneling effects, and so on. Moreover, they play an essential role in electrochemical analysis and have various applications [5, 6]. Since carbon nanomaterial's discovery, they have been at the forefront of nanotechnology. Graphene (GR), carbon nanotubes (CNTs), and fullerenes (C60) are the most common carbon nanomaterials. Carbon nanoparticles' usage in electrochemical sensors has shown promise for future investigation and application [7, 8].

Carbon nanotubes and graphene are widely used as the electrode material modifier for detecting electrochemicals in flavonoids as a result of their fine features, such as large surface area, small size, high sensitivity, and quick electron transfer [9]. Graphene is a two-dimensional carbon nanomaterial with honeycomb-shape, while Carbon nanotubes can be in forms of single-walled carbon nanotubes (SWCNTs), double-walled carbon nanotubes (DWCNTs), and multi-walled carbon nanotubes (MWCNTs) composing of graphene layers wrapped into nanometer-sized sheets. In a study conducted by Dar et al., a modified MWCNT carbon paste electrode (CPE) was used for the determination of atropine (an alkaloid subclass) in seeds and leaves of *Datura stramonium*, an Indian traditional medicine [10]. In another study, ascorbic acid (AA) and caffeine (CAF) concentration in different matrixes, such as coffee, tea leaves, cold drink (Mountain Dew), pharmaceutical preparations, and urine samples were studied. Determination was achieved using glassy carbon electrode (GCE) modified with multiwall carbon nanotubes (MWCNT) with square wave voltammetry (SWV) method [11]. Lin et al. used SWCNT carbon nanomaterials as a carbon paste electrode modifier for the simultaneous measurement of two effective substances, including quercetin and rutin, in real samples of diluted blood serums [12]. Ghoreishi et al. used SWCNT carbon nanomaterials as a carbon paste electrode modifier for the simultaneous measurement of two effective substances, including ellagic acid and gallic acid, in *Punica granatum*, *Myrtus communis*, and Itriphal plants [13]. Ghoreishi et al. used SWCNT carbon nanomaterials as a carbon paste electrode modifier for the simultaneous measurement of two effective substances, including ellagic acid and tannic acid, in *Punica granatum* and *Quercus infectoria* plants [14].

4.2.2 Carbon Nanocomposite–Modified Electrodes

Graphene and carbon nanotubes are the two basic material in almost every electrochemically modified electrode in combination with other materials. Composites can work together to create superior electrochemical catalytic impact as a result of electrode surface change. Two or even three distinct components combine with each other and form carbon nanoparticles. To make a carbon-based nanocomposite for natural compounds detection, many materials are mixed with carbon nanoparticles. For chrysin detection, Li et al. added zirconium dioxide (ZrO_2) nanoparticles and single-walled carbon nanotubes (SWCNTs) to glassy carbon electrodes in existence of sodium dodecyl sulfate (SDS) as surfactant [15]. In another study of rutin detection, a graphene-modified glassy carbon electrode modified with polydiallyldimethylammonium chloride was used by Miao et al. [16]. Okutan et al. have modified glassy carbon electrodes (GCE) by adding graphene-based zirconia nanocomposite for caffeine detection in various beverages [17]. In another investigation, platinum (Pt) nanoparticles decorating a biomass-porous carbon nanocomposite–modified electrode was employed for the determination of luteolin conducted by Liu et al. [18].

Hou et al. used carbon nanocomposite chitosan/reduced graphene oxide aerogel with dispersed ZrO_2 (ZrO_2/CS/rGOA) as a carbon glass electrode modifier to measure the active ingredient luteolin in real samples, including peach juice and red wine [19]. Xu et al. used MXene/ZIF-67/CNTs carbon nanocomposite as a carbon glass electrode modifier to measure the active ingredient luteolin in a real grape juice sample [20]. Ansari and Arvand used carbon nanocomposite of cobalt ferrite magnetic electrospun nanofibers (NFs) and graphene oxide (GO) ($CoFe_2O_4$/GO) as a carbon paste electrode modifier to measure the active ingredient rutin in real samples including red apple, lime, lemon, and orange juices [20]. Şenocak et al. used SWCNT-subphthalocyanine(CS) carbon nanocomposite as a carbon glass electrode modifier to measure the active ingredient catechin in tea plant [21].

4.2.3 Metal Nanoparticles Application in Modified Electrodes

Platinum (Pt), gold (Au), silver (Ag), palladium (Pd), copper (Cu), and other metals can also be used as nanomaterials. Due to nanomaterial features, metal nanoparticles have specific electrochemical characteristics. A composite-modified ionic liquid carbon paste electrode was prepared using AuNPs, PtNPs, and biomass-derived porous carbon by Liu et al., and the sensitivity and electrochemical behavior of quercetin in ginkgo biloba leaves were investigated. The result showed a broad application of these two nanoparticles in flavonoid detection. To form nanocomposite-modified chemical electrodes, metal nanoparticles doped in carbon nanomaterials have been utilized and are wildly employed. The combination of these materials has a synergistic impact on the electrode detection selectivity and sensitivity for analytes. In a study, Mahanthappa et al. have modified a carbon paste electrode using copper sulfide nanoparticles. The CuS NPs–modified carbon paste electrode (CuS NPs MCPE) showed significant activity toward caffeine determination by voltammetric methods [22]. Fekry et al. employed nano-cobalt (II, III) oxide–modified carbon paste (NCOMCP) sensor for the determination of caffeine in aqueous and micellar media [23].

4.2.4 Ionic Liquid–Modified Electrodes

A liquid consists of both cations and anions, named ionic liquid. Based on the anions and cations that are used, various types of ionic liquids would be made. Ionic liquids have advantages that make them suitable for electrochemistry analysis, such as wide potential window, high electrical conductivity, stability, and non-volatility. Ionic liquid can be applied in combination with other carbon materials or dropwise to the electrode surface to make a composite-modified electrode. For the determination of rosmarinic acid using square-wave voltammetry, a biosensor based on laccase and an ionic liquid were constructed by Franzoi et al. [24]. Sun et al. have studied electrochemical characteristics of rutin with a pyridinium-type ionic liquid–modified carbon paste electrode (IL-CPE) [25].

4.2.5 Molecularly Imprinted Polymer–Based Electrochemical Sensors

By utilizing molecular imprinting processes, MIPs (molecularly imprinted polymers) are made. Molecular imprinting technology (MIT) is a modern technique; the process involves a covalent or non-covalent connection of a target molecule to a functional monomer, and later, the target molecule is removed from the polymer to form polymer-containing steric holes. The cavity, has the capacity to bind and detect the target molecule precisely due to its molecular structure and functional group arrangement. For better understanding, the connection of the specific identification of target molecules by MIP, the interaction of lock and key, antigen and antibody, enzyme and 2ubstrate, can be a good graphic illustration. Exact identification of target compounds, extended shelf life, structure stability, and good selectivity are the main characteristics of MIP. MIP electrochemical sensors have been used in the analytical detection of flavonoids as a result of their excellent sensitivity, quick reaction time, and being cost-efficient. Comino and colleagues have employed a voltammetric sensor for the determination of catechol in wine samples. The sensor was based on a molecularly imprinted chitosan film electrodeposited on a boron-doped diamond (BDD) electrode and represented high reproducibility in complex matrixes [26]. In another work, Yang et al. have developed gallic acid–sensing in edible plants, using a potentiometric sensor based on molecular imprinting polymer [27].

Pierini et al. have used screen-printed electrodes as a modified electrode to measure the active ingredient taxifolin in the natural composition of edible peanut oils [28]. Munteanu et al. used screen-printed carbon electrode modified with graphene and gold nanoparticles as an electrode for measuring the active ingredient chlorogenic acid in real green coffee samples [29]. Guiberteau et al. have used graphene–modified screen-printed electrode as a modified electrode to measure two active substances, including ellagic acid and gallic acid in real sample of cork boiling waters [30].

4.3 TYPES OF VOLTAMMETRY METHODS

4.3.1 Cyclic Voltammetry (CV)

In the cyclic voltammetry method, like the linear scan voltammetry method, the application of the potential is in the form of a scan, with the difference that the potential in a range is scanned back and forth, and the recording of the current corresponding to each potential is drawn as a curve. In potential scanning methods, including cyclic voltammetry, with increasing speed of potential application, a larger number of sample ions undergo electrochemical reaction in the transducer. The main feature of the cyclic voltammetry technique is that the current can be changed in it and depends on the speed of the potential scan; according to this feature, one can be aware of the role of surface absorption, penetration, and accompanying chemical reactions. So in the cyclic voltammetry method, by increasing the speed of potential application, more samples are subjected to electrochemical reaction and the current increases, but this increase in current is observed only as long as the system is under penetration control. From the limit current onwards, the system is under the control of concentration polarization, and in this case, there is no more species that undergoes electrochemical reaction; as a result, the thickness of the penetration layer increases, and the amount of current decreases. This method is considered a powerful tool in electrochemical studies, in which valuable information about the electrochemical processes of the system and the determination of kinetic parameters can be obtained by scanning the potential and checking the current changes. For example, among the applications of cyclic voltammetry, we can mention the investigation of the corrosion behavior of metals, the oxidation-reduction properties of materials, and the determination of the penetration coefficient. Cyclic voltammetry can also determine the reversibility and irreversibility of the electrochemical system. In this method, according to the way of applying the potential, it is possible to determine which of the electrochemical half-reactions of oxidation and reduction should be performed first. Cyclic voltammetry is also a powerful technique in determining the mechanism of reactions and the number of exchanged electrons, determining kinetic and thermodynamic parameters, and determining electrochemical couples [31]. How to apply the potential with respect to time in cyclic voltammetry and the resulting voltammogram curve are shown in Figure 4.1, respectively.

4.3.2 Linear Sweep Voltammetry (LSV)

The linear scanning voltammetry method is a type of voltammetry in which the potential is swept in a constant potential range at a certain speed and in a direction. In the sense that in applying the potential in this method, a potential sweep is used instead of a potential step; for this reason, in the linear-sweep voltammetry method, due to the permanent presence of the capacitive current, the current never becomes zero. In this method, the drawn curve belongs to the recording of the current corresponding to each potential. The intensity of recorded currents depends on the rate of charge transfer reaction, the chemical activity of electroactive species, the speed of potential scanning, etc. This type of voltammetry is used to identify the potential of electrochemical reactions and the amount of charge exchanged, draw current–potential curves, determine the concentration of unknown species, etc. The detection limit of this method is usually between 10^{-5} and 10^{-7} M, which depends on the type of applied potential and current sampling method [31]. Also, by drawing the TOEFL curve (potential diagram according to the logarithm of current) and mathematical relations 1 to 3, it is possible to calculate the charge transfer coefficient (α) and the

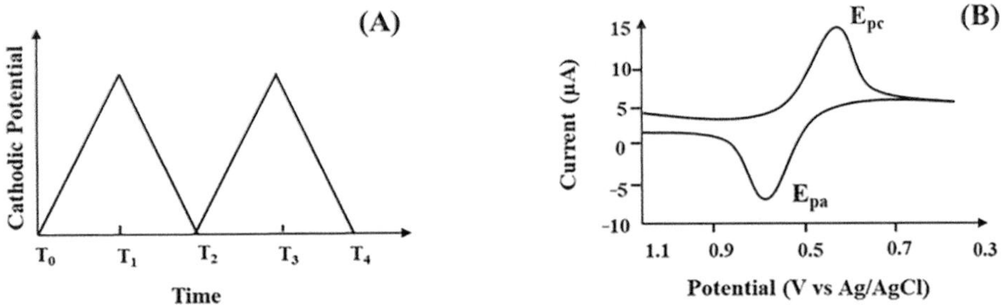

FIGURE 4.1 (a) How to apply potential to time in cyclic voltammetry. (b) Resulting voltammogram curve.

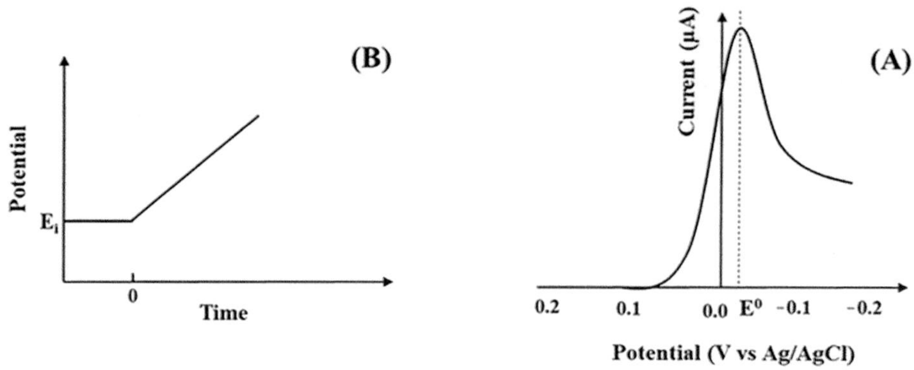

FIGURE 4.2 (a) Linear scan voltammogram. (b) How to apply the potential.

exchanged current density (i_0). Also, Figure 4.2 shows how to apply the potential and the resulting voltammogram.

$$\eta = a + b \log i$$

$$a = -2.3\ RT \log\left(-i_0\right)/n_a\left(1\text{-}a\right)$$

$$b = 2.3\ RT/n_a\left(1\text{-}a\right)F$$

4.3.3 Chronocoulometry

In the chronocoulometry method, kinetic constants are obtained from the analysis of load data as a function of useful time. Through the integration of Cottrell's equation, it can be seen that the charge changes with respect to the square root of time in the form of the equation $Q = 2nFACD^{1/2}\, t^{1/2}\, \pi^{-1/2}$. This equation is used in a wide time range, and it can be used to calculate the diffusion coefficient of the species [31]. How to apply the potential in the chronocoulometry method and the resulting chronocoulogram are shown in Figure 4.3. To obtain the diffusion coefficient using this method, first, the Q diagram is drawn in terms of $t^{-1/2}$ from the resulting chronocoulograms for different concentrations of the electroactive species. Finally, by plotting the slope of the obtained curves according to the concentration of the electroactive species, the penetration coefficient is obtained.

4.3.4 Differential Pulse Voltammetry (DPV)

All kinds of pulsating voltammetry methods are used in analytical cases due to their high sensitivity. One of the most important of them is the differential pulse voltammetry method, in which pulses with a constant amplitude (between 1 and 100 mV) are applied to the gradient potential of the background, and the current intensity

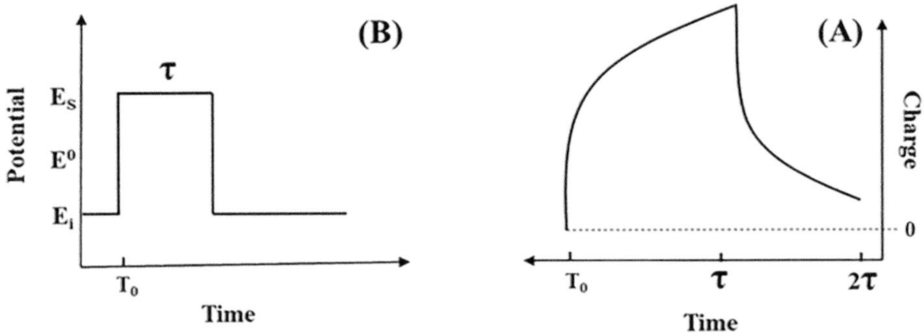

FIGURE 4.3 (a) How to apply the potential in the chronocoulometry method. (b) The resulting chronocoulogram.

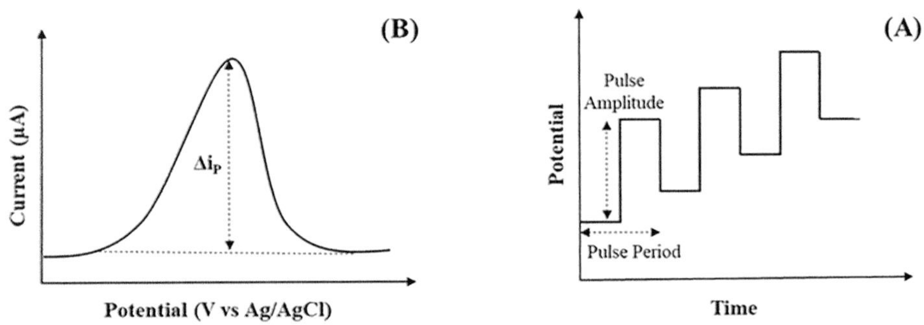

FIGURE 4.4 (a) How to apply the potential over time in DPV. (b) The differential pulse voltammogram curve of an electroactive species.

before applying the pulse and at the end of the pulse life is sampled, and the changes in the difference between the two currents is recorded according to the gradient potential applied to the electrode. The shape of the differential pulse voltammograms is a peak with a stepped appearance, and is suitable for analytical purposes, and the detection limit of this method is around 10^{-8} M [31]. A view of the sweep in differential pulse voltammetry, including the lifetime of each pulse, the distance between two consecutive pulses and sampling points, as well as the resulting voltammogram curve, is shown in Figure 4.4.

4.3.5 Electrochemical Impedance Spectroscopy (EIS)

In alternating voltage systems where the current and the potential change sinusoidally have a phase difference of $\pi/2$ with respect to each other, the alternating resistance caused by this system is called impedance. The impedance method is used to check the kinetics of electron transfer reactions; therefore, according to Warberg's studies, the impedance should be checked at high frequencies, where the system

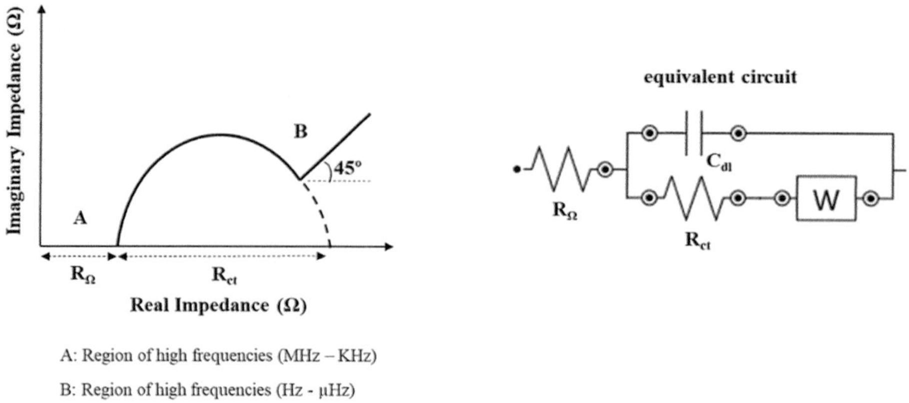

FIGURE 4.5 An example of an EIS spectrum and its equivalent circuit.

is under kinetic control. At low frequencies, the electroactive species has a lot of opportunity to be oxidized and reduced, and electrolysis occurs on the surface of the cathode electrode. In this case, the thickness of the penetration layer increases, and the system will be under penetration control. So to check the impedance, move from high to low frequencies so that the kinetics of electron transfer can be checked. Also, in the investigation of kinetics with the help of impedance, a very small overpotential should be applied to the system compared to the inherent potential (Nernst relation potential); otherwise, the system will have a twist in the current–potential sinusoidal response. Electron transfer reactions are investigated by using changes in Rct (resistance of charge transfer or polarization). During the application of the potential, an electric double layer is formed at the solution–metal interface, which is represented by C_{dl}. Also, R_Ω (uncompensated resistance) is formed due to the resistance of the solution between the working electrode and the reference electrode. The existence of these two components makes the calculation work more difficult. Therefore, the mixed-plane impedance analysis method will be used. Impedance graphs are drawn as real impedance (Z′) based on imaginary impedance (Z″). Then a comparison is made between the considered electrochemical cell and an equivalent electric circuit, including capacitors and resistors [31]. Figure 4.5 shows the simplest type of equivalent circuit, along with an EIS spectrum.

4.4 REACTIVE FUNCTIONAL GROUPS IN NATURAL AND HERBAL COMPOUNDS

Electroanalysis is based on recording electrochemical signals from plants, pigments, and cultures, and the signal can be considered a fingerprint [32, 33]. The only reactions that can be detected by electrochemistry are reactions with redox kinetics and biodetection steps, including current or potential impedance or non-electrochemical properties, including van der Waal interactions, fluctuations in the electrical responses due to structural changes, matrix intercalation caused by electric fields, mass transfer beyond the electrical double layer [34].

Natural products are synthesized chemical substances with animal, microbial, and plant origin. metabolites are divided in to primary and secondary metabolites. Those with role in plants' respiratory and photosynthetic metabolism are named as primary. Proteins, carbohydrates, lipids, and nucleic acids are the most common primary metabolites [35]. Secondary metabolites are organic compounds with important functions in plants, such as defense system against insect repellence, that help in pollination and plant reproduction [36]. Secondary metabolites, based on their chemical structure, are classified in three major classes: phenolic compounds, nitrogen-containing compounds, and terpenoids [37].

4.4.1 Phenolic Compounds

Phenolic compounds are a significant group of antioxidants with at least one phenolic ring in their structure. This class comprises phenolic acids and phenolic alcohols with one phenolic ring, and polyphenols which have more than one phenolic unit. All polyphenols have the same redox behavior, electrochemical oxidation occurring at the –OH groups, and influenced by chemical substituents linked to the aromatic rings (e.g., $-OCH_3$, sugar, etc.). The pH of the environment is the most important factor with direct effect on the polyphenols' redox behavior, antioxidant capacity, and oxidation product formation [38].

4.4.1.1 Determination of Phenolic Compounds with One Phenolic Ring

One of the most significant natural antioxidant groups are phenolic acids, consisting of hydroxycinnamic and hydroxybenzoic acids and their derivatives. hydroxycinnamic acids are cinnamic acid derivatives, while the hydroxybenzoic acids are benzoic acid derivatives [39].

They are all around in plant: seeds, fruits, roots, and food products, such as grapes, beer, tea, coffee, officinal herbs, honey, fruit juices. Phenolic acids are considered powerful antioxidants, and their antioxidant capacity depends on the number and position of the-OH moieties present on the benzene ring.

1. *Hydroxybenzoic acids.* Gallic acid (GA) is one of the most important antioxidants belonging to hydroxybenzoic acids.
    - *Gallic acid (GA).* GA can be found in grapes, tea leaves, and red fruits, such as berries, strawberries, and so on, and as well as in many other officinal herbs and in drinks and beverages, such as teas and red wine. GA was regarded as effective against the development of different diseases, such as cancer and cardiovascular and neurodegenerative diseases, due to its antioxidant capacity. Gopal et al. investigated the electrochemical redox behavior of GA by using CV technique at different electrodes. GA has shown two irreversible oxidation peaks corresponding to the oxidation of the catechol group (at lower potentials) and the oxidation of other phenolic groups (at higher potentials) present in the GA [40].
2. *Hydroxycinnamic acids.* Hydroxycinnamic acids are another phenolic acid subgroup, synthesized by a metabolic route named shikimic pathway, generally operating in plants, fungi, and algae (4,76–78). Here we will discuss

some of hydroxycinnamic acid derivatives, including caffeic acid, rosemaric acid, ferulic acid, and chlorogenic acid.

- *Caffeic Acid (CA).* CA is a type of polyphenol that is widely distributed in plants as glycosides, esters, and freely. CA naturally exists in many agricultural products, such as vegetables, fruits, grains, seasonings and spices, wine, olive oil, alcoholic and non-alcoholic beverages, dark chocolate, and coffee [41]. CA has been used as an anticancer, antibacterial, antiallergy, immune system regulator, and for its antiviral activity. Furthermore, CA can prevent asthma and is consumed as a compound with some anti-HIV properties [42]. The redox mechanism of CA which provides its sensing with electrochemical sensors is shown in what follows. It can be observed that the redox process is reversible and involves the exchange of two protons and two electrons. CA is electrochemically oxidized to the o-quinone derivative, (E)-3–(3,4-dioxocyclohexa-1,5-dien-1-yl) acrylic acid. The process is reversible: The o-quinone derivative is then reduced to caffeic acid. when the detection technique is the cyclic voltammetry, this reversible process is well evidenced [43].
- *Rosmaric Acid (RA).* Although RA was separated for the first time from rosemary (*Rosmarinus officinalis* L.) in 1958, its chemical synthesis was accomplished in 1991. From a structural point of view, RA contains an ester bond between an essential aromatic α-amino acid and a nonessential one, namely, phenylalanine and tyrosine.
  RA is present in many plants, herbs, spices, vegetables, and fruits. It exists in culinaryand medicinal plants as a secondary metabolite. Therefore, here are some important sources of RA: rosemary, mint, spearmint, lemon balm, perilla, salvia, basil, sage, marjoram, different species of thyme, oregano, savory, tamarillo. Ferns, hornworts, and sea grass must also be mentioned as sources of this polyphenolic compound. It is worth emphasizing that RA is characteristic for thyme honey.
  RA has multiple biological and pharmaceutical activities: antioxidant, anti-inflammatory, antimutagenic, antibacterial, antitumor, hepatoprotective,

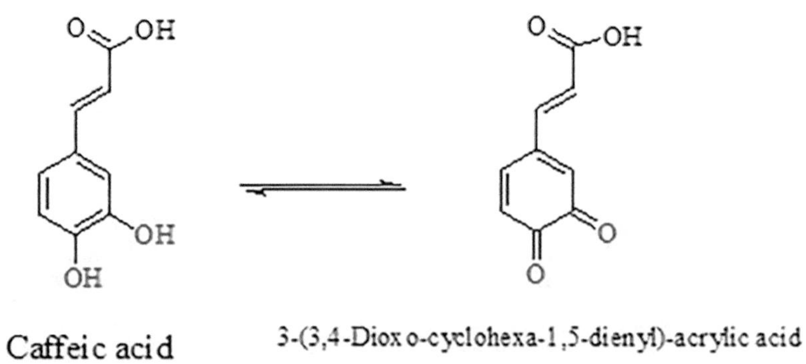

FIGURE 4.6 The redox mechanism of CA.

and antihepatitis, cardioprotective, antiviral, anticarcinogenic, anticonvulsant, immunomodulatory, antimetastatic, antiangiogenic, astringent, and antiallergic. It also helps the fertilization capacity increase and in cognitive efficiency enhancement (neuroprotective activity).
RA, a naturally occurring antioxidant, is an ester of caffeic acid (CAFA) and 3,4-dihydroxyphenyl lactic acid (DHPLA). It comprises two catechol moieties, thus having two pairs of ortho-hydroxyl groups grafted on two phenolic rings, which can be easily oxidized, being thus responsible for RA electroactivity [44].

- *Ferulic acid (FA, 3-methoxy-4-hydroxy cinnamic acid)* is used in many traditional Chinese medicines; mostly found in *Pinellia ternata*, *Angelica sinensis*, Ligusticum chuanxiong, propolis, and Semen Ziziphi Spinosae [45, 46]. Recently, studies on the pharmacological activities of ferulic acid have shown that ferulic acid and its derivatives have many biological activities, such as antioxidant, antithrombotic, anti-inflammatory, antineoplastic, antimicrobial, hypolipidemic, and immunomodulatory functions [47]. Studies on ferulic acid redox behavior suggest that ferulic acid molecules undergo a first electron transfer, leading to the formation of phenoxy radicals, for which the electronic charge distribution can be represented by four mesomeric forms A, B, C, and D (scheme 1).

The radical D would be the most probable as a reason of the electronic effects of methoxy, carboxyvinyl, and oxo groups. Caffeic acid formation could be explained assuming that the radical D undergoes a second electron transfer, leading to the carbocation *D'*, which gives simultaneously the 3,4-dioxocinnamic acid (I, scheme 4) and a methanol molecule by hydrolysis.

During the reverse scan, the 3,4-dioxocinnamic acid is reduced to caffeic acid (J, scheme 4) at the potential of the reversible peak I4. Furthermore, the proportions of ferulic acid oxidation products formed via the intermediary of the less probable forms B and C would be insignificant. The oxidation of radical B leads to the formation of methoxyparabenzoquinone (E, scheme 2), which is reduced to methoxyhydroquinone (J, scheme 4) at the potential of the reversible peak I2. The formation of 3,4-dihydroxy-5-methoxycinnamic acid (G, scheme 3), stemming from the oxidation of the radical C, was not recognized given that it is not commercial [48].

- *Chlorogenic acids (CGAs)* are one of the sub-branches of the large group of polyphenolic compounds. CGAs include different esters (mono-, di-, tri-, and mixed esters) formed between (-)-quinic acid and certain hydroxycinnamic acids. The most common ones are caffeic, ferulic, and p-coumaric acids. They may be subdivided into more subgroups due to the type of ester substituent (e.g., caffeoylquinic acids, CQAs; feruloylquinic acids, FQAs; p-coumaroylquinic acids, p-CoQAs; different mixed esters, e.g. caffeoyl-feryloylquinic acids, CFQAs, etc.). Individual acyl residue position (e.g., 5-caffeoylquinic acid, 5-CQA) and the number of ester substituent (monoacylquinic acids, diacylquinic acids, or triacylquinic acids) are essential parameters [49, 50].

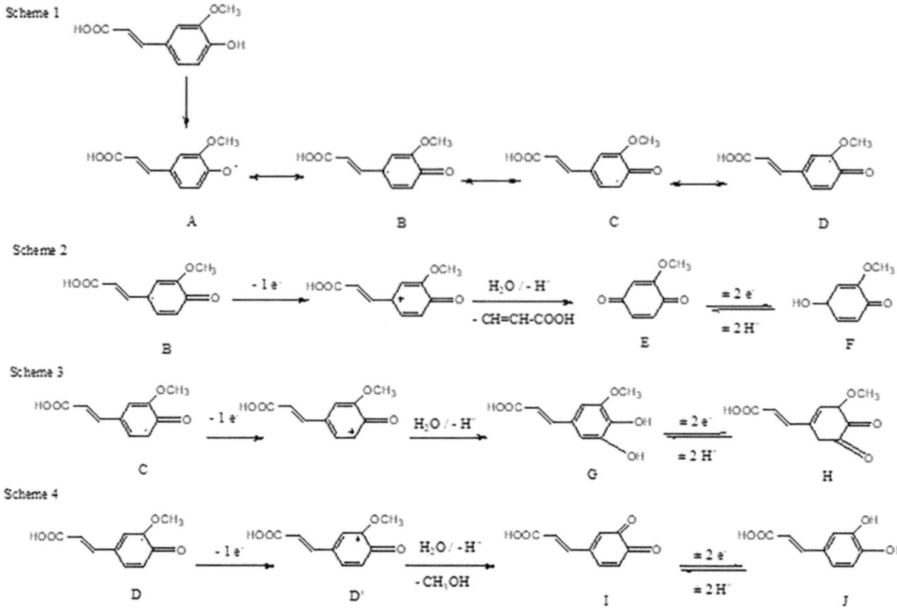

FIGURE 4.7 The redox behavior of ferulic acid.

It has been proven that having a diet rich in CGAs (particulary through coffee beverage consumption) has many pharmacological properties, such as positive effects on many types of diseases, like type 2 diabetes, some neurodegenerative diseases (Alzheimer's and Parkinson's), cardiovascular and liver health, type of cancer (like prostate, pancreas, bladder, breast, colon), etc. [51]. Coffee is a rich source of different CGAs, for instance, green coffee bean, which has about 70 isomers, although in roasted bean it increases to over 200 derivates [50].

Tomac et al. had studied on the electrochemical properties of all CGA derivatives and had shown that it strongly depends on CGAs chemical structure and electronic properties, particularly on the electron-donating effect of –OH and –CH=CH– groups, and a strong electron-withdrawing effect of ester (-COOR) group presented in their structure [52].

4.4.1.2 Determination of Phenolic Compounds with More Than One Phenolic Ring (Polyphenols)

There are various categories for phenolic compounds, the most common being phenolic acids, phenolic alcohols, lignans, stilbenes, and flavonoids. Flavonoids are also divided into six subclasses, namely, flavonols, flavones, isoflavones, flavanones, anthocyanidins, and flavanols (catechins and proanthocyanidins) [53]. The diversity of flavonoids is due to their structural class, functional substitutions, conjugation and degree of polymerization, and degree of hydroxylation.

Flavonoids have three phenol ring structures, named A (benzene ring), B (second phenyl ring), and C ring. this is common to all flavonoids and is characterized by

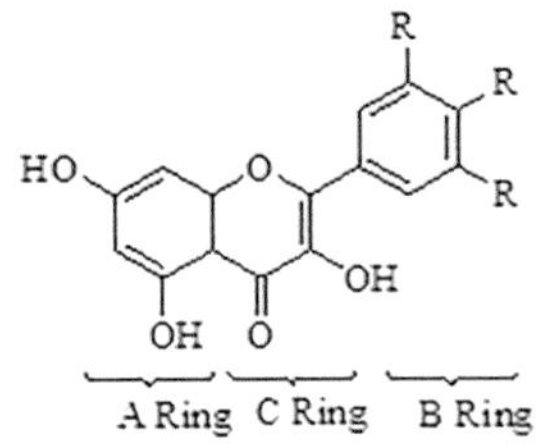

FIGURE 4.8 The basic structure of flavonoids.

C6–C3–C6 carbon skeleton [54]. The situation of hydroxyl group strictly correlates antioxidant activity and creates different derivations [55].

4.4.2 Electrochemical Sensing of Various Flavonoids

The redox property for measuring flavonoids is caused by electroactive stable phenoxyl radicals produced during anodic oxidation. Phenolic groups on ring B and ring A, with their electron/proton donor ability, have notable antioxidant activity in comparison to ring C.

Adsorption-controlled or sole diffusion reaction at the electrode–electrolyte interface enables the selective electrochemical measurement of flavonoids.

For the anodic oxidation of flavonoids, the electron transfer mechanism at the interface works according to a pH-dependent single-electron proton, or multi-electron proton transfer mechanism, and it could be a reversible, quasireversible, and irreversible process. In what follows we discuss some flavonoids subgroup structure, redox behavior, sources and usage in human's diet [56].

4.4.2.1 Luteolin

Luteolin (30,40,5,7-tetrahydroxyflavone) $\left(C_{15}H_{10}O_6\right)$ is an essential medicinal flavonoid under the subclass flavone present in a wide range of vegetables and fruits, such as carrots, peppers, thyme, rosemary, celery, olive oil, oregano, and peppermint [57]. Luteolin's pharmacological activities include anti-inflammatory, antitumor, antibacterial, antiallergic, antiulcer, cataract prevention, and antiviral activities and also act as an antioxidant or a pro-oxidant. Because of its high potential in treating diseases, luteolin was used in Chinese traditional medicine for treating inflammatory disorders, cough, cardiovascular diseases, and so on. Luteolin could be quantified with various methods, among them electrochemical methods, which have gained much attention due to the presence of an electroactive catechol group on the B ring (30,40-dihydroxyl) of luteolin, which produces a clear redox peak pair [58–60].

4.4.2.2 Quercetin

One of the flavonols' subclass is quercetin (5,7,30,40 -flavon-3-ol) $\left(C_{15}H_{10}O_7\right)$, one of the major dietary antioxidants found in onions, apples, kale, tea, wine, as well as in numerous food supplements. Usually, a sugar group like glucose, galactose, xylose, rutinose, or rhamnose is attached to one of the quercetin hydroxyl groups and forms a glycoside [9, 14, 35, 36]. The beneficial health effects of this flavonoid are a cause of its highly reactive radical scavenging activity toward the peroxynitrite and the hydroxyl radicals. This natural compound has therapeutic effect, such as protection against pulmonary diseases, cardiovascular diseases, aging, certain types of cancer, osteoporosis, and so on.

Quercetin could be effective in cancer therapy and be used as an alternative to high-cost conventional cancer treatments, hence has the ability of anti-proliferation, cell cycle arrest, and apoptosis [11]. In quercetin, of the five hydroxyl groups present, the 3',4' -dihydroxy electron-donating (catechol) groups of the B ring show the highest level of electroactivity and its oxidation peak results from a two-electron, two-proton transfer reaction to produce the corresponding o-quinone (3',4' -diquinone) product. [37]

When the basal levels of oxidative stress and inflammation are high, the antioxidant and anti-inflammatory effects of quercetin are found to be more pronounced.

Reactive oxygen species' (ROS) scavenging activities of quercetin accelerate the formation of oxidation products including orthoquinone, which and have genotoxic effects and are highly reactive [14]. Some research result indicates the cytotoxic, mutagenic, and genotoxic properties of quercetin [11–13].

4.4.2.3 Rutin

Rutin (3, 30, 40, 5, 7-pentahydroxyflavone-3-rhamnoglucoside), also known as vitamin P $\left(C_{27}H_{30}O_{16}\right)$, rutoside or quercetin-3-O-rutinoside belongs to the subclass of flavonol. Rutin is a citrus flavonoid glycoside and can be found in many plants, including tea, apple, passion flower, and buckwheat, and have pharmacological, nutraceutical, and chemotherapeutic effects, such as anticarcinogenic, anti-inflammatory, antiasthmatic, cardioprotective, cytoprotective, hepatoprotective, nephroprotective, neuroprotective, and vasoprotective activities. Due to rutin's high radical scavenging properties, it acts as an antioxidant and metal chelating property, inhibiting metal ion–induced peroxidations. Rutin is protective against UV radiation, Alzheimer's, diabetics, neurodegenerative disorders, hyperkinetic movement disorder, stroke, hypertension, etc., and also promotes thyroid uptake and neural crest cell survival. However, its low stability to processing conditions, low water stability, and possible interactions with other components create limitation in the usage of rutin into functional foods, food supplements, or drugs. Although stabilized rutin can be used as a part of food packaging. Electrochemical studies on rutin determination, its health benefits and functional properties have been widely studied. mainly, the complex process of electro-oxidation of rutin proceeds by a cascade mechanism and involves the 4-hydroxyl groups. The 30,40 -dihydroxyl catechol group first oxidizes by a reversible two-electron–two-proton reaction, and then the 5,7-dihydroxyl group undergoes irreversible oxidation.

These mechanisms highly depend on pH-d and is followed by either diffusion-controlled or adsorption-controlled process or a combination of both. oxidation products can accumulate and block electrode surface [6, 54, 55].

4.4.2.4 Morin

Morin is a bioactive flavonoid (2',3,4',5,7-pentahydroxyflavone) $(C_{15}H_{10}O_7)$ belonging to the subclass of flavonols and found in coffee, tea, old fustic, yellow Brazil wood, Osage orange, leaves of Psidium guajava (common guava), and so many fruits and vegetables. Morin shows numerous pharmacological activities [61, 62]. Morin has antioxidant activity and can protect human cells, including myocytes, endothelial cells, and hepatocytes, against the in situ generated oxy radicals. Reports indicate that morin may safeguard the cardiovascular tissues against toxicity and damage brought about by free radicals in cancer patients due to the use of doxorubicin or mitomycin C in antitumor drugs [63, 64]. For investigating physicochemical parameters, antioxidative activities, and oxidation reaction mechanisms of morin, electrochemical methods can be used. The radical scavenging abilities and the electron/proton donating of morin are caused by the hydroxyl moiety of the B ring. By using voltammetry methods, electro-oxidation mechanisms of solid, insoluble morin hydrate on GCE surface in aqueous media were studied. The hydroxyl groups of the three aromatic rings of morin undergo pH-dependent oxidation reactions in sequential steps, the electron-donating 2',4'dihydroxy group of B ring being the first to oxidize reversibly at nether positive potentials with high peak current by a one-electron one proton mechanism, followed by hydroxyl group at the 3-position of the C ring. Oxidation of hydroxyl part at higher positive potentials undergoes an irreversible oxidation mechanism [56, 65].

4.4.2.5 Catechin

Catechin (flavan-3-ol) $(C_{15}H_{14}O_6)$, mostly present in fruits, wine, and cocoa products. (+)-Catechin is the one most present in food; however, chocolate, mainly contains (-)-Catechin, in addition to (-)-epicatechin flavanol. Previous studies indicate chocolate consumption shows less catechin bioavailability compared to other foods. Due to differential absorption in the small intestine, a differential affinity for glucuronosyl transferase of the enzyme sulfotransferase, or transport proteins, and also possible isomerization of some (-)-catechin to (+)-catechin in the intestinal enterocytes, concentrations of the (-)-Catechin in the plasma is way more less than (+)-Catechin [66]. Catechin is a natural radical scavenger with two pharmacophores, the resorcinol moiety, which is the A ring, and the catechol moiety, responsible for its antioxidant activities, which is the B ring. Catechin has significant pharmacological activity such as protective effects against inflammation, lipid proliferation, LDL oxidation, diabetes mellitus, neurodegenerative disorders, and coronary heart diseases, decelerating tumor growth and suppressing the breakage of DNA strands [67]. Consecutive steps of Catechin oxidation starts with reversibly oxidation of electron-donating 3,4-dihydroxyl catechol groups, at very low positive potentials, followed by irreversible reaction of the hydroxyl groups of resorcinol moiety. The oxidation process is pH-dependent, and (+) catechin is strongly adsorbed on the surface of the electrode and the electro-inactive final oxidation product accumulates and blocks the surface [56].

4.4.2.6 Hesperidin

Hesperidin (hesperetin-7-rutinoside) $(C_{28}H_{34}O_{15})$, found in citrus fruits, such as lemon, oranges and satsuma mandarin is a major flavanone with antioxidant, antiallergenic, antimicrobial, anti-hypotensive, and vasodilatory impacts. It has protective properties against atherosclerosis and cancer and reduces blood triglycerides and cholesterol. The antioxidant properties and radical scavenging of thse flavonoids, which are attributed to its chemical structure involving the electron-donating 3"-hydroxy,4"-o-methoxy moiety of the B ring, C4' -C8' double bond conjugated with the 4-carbonyl group of C ring, and 5-hydroxyl group and C5–C6 double bond of the A ring [56]. Hesperidin will be absorbed in its glycoside form in the intestine and after 3 hours of ingestion, hesperitin, its aglycone appears in the plasma reaching a maximum concentration in about 5–7 hours. Glucuronides (87%) and sulfo-glucuronides (13%) are the circulating forms of hesperetin, and hesperidin is almost completely excreted through urine after 24 hours irrespective of the dose [56, 68].

4.4.2.7 Taxifolin

Taxifolin or dihydroquercetin (3,5,7,3,4-pentahydroxy flavanone) $(C_{15}H_{12}O_7)$ is a flavanonol present in milk thistle, onion, Douglas fir bark, and French maritime pine bark. It can be used as an antioxidant-rich functional food or in dietary supplements. Taxifolin, both cis- and trans-forms, are soluble in polar solvents as well as water–alcohol solutions and exhibits a wide range of pharmacological and therapeutic activities and protects against inflammation, microbial infections, oxidative stress, tumors, toxoplasmosis, cardiovascular disorders, Alzheimer's and liver dysfunctions. (+)-trans-dihydroquercetin more readily oxidizes, donating its hydrogen atoms to form quercetin (2–(3,4-dihydroxyphenyl)-3,5,7-trihydroxy-4H-chromen-4-one) as the oxidation product. the 5- and 7-OH moieties with 4-oxo function in the A and C rings, along with the stability provided by the O-dihydroxy system in the B ring, possess Taxifolin radical scavenging activities [69]. Taxifolin can be used as lipid oxidation inhibitor in pharmaceutical and food products to maintain their nutritional quality as well as help in prolonging their shelf life [70].

4.4.2.8 Apigenin

Apigenin (5,7,40 -trihydroxyflavone) is a flavone mostly found in parsley, celery, and chamomile. The compound has very low intrinsic toxicity; hence, it can be used as the chemo-preventive and anti-proliferative activities against different tumoral lesions [71]. Various studies on apigenin have paved the way for finding new cancer treatment drugs because apigenin has effect on inducing apoptosis in cancer cell lines via the hindering different enzymes systems. Apigenin antioxidant mechanism is responsible for its biological effects, although as the redox agent, it can play as an oxidant having a negative impact. Knowing apigenin oxidation mechanism is critical for understanding its biochemical properties. Electrochemical methods were employed to study the redox behavior of several flavonoids, their metal chelating, reactive oxygen species scavenging, as well as DNA interaction features. The research indicates apigenin electron-donating abilities, and it was demonstrated that the oxidation potentials can be related to the antioxidant power and the hydroxyl groups in the B-ring are more easily oxidized than those in the A-ring. Generally, compounds containing a cathecol group can be reversibly oxidized and the lower low oxidation potential, the higher antioxidant capacity. Oxidation mechanism could be affected in the absence of cathecol group. Popa et al. employed cyclic, differential pulse and square wave voltammetry for investigating apigenin with a glassy carbon electrode. Results have shown that the oxidation process is pH-dependent and irreversible, happening at a cascade mechanism, with transferring one electron and one proton from the hydroxyl groups attached to their structures in each step [72].

4.4.2.9 Genistein

Genistein (4,5,7-trihydroxy isoflavones) is a typical phytoestrogen and an important member of the isoflavones family. Being a natural product of leguminous plants, genistein exhibits multitudinous biological properties and pharmacological effects [61]. The electrochemical oxidation mechanisms of isoflavones genistein and biochanin A, compounds with anticancer properties and inhibitors of protein tyrosine kinases, were investigated at a glassy carbon electrode by cyclic, differential pulse and square wave voltammetry. It has been shown that in acid, neutral and mild alkaline electrolytes, each hydroxyl group of genistein undergo; irreversible, pH-dependent oxidation with the transfer of one electron and one proton. For strong alkaline media, two pH-independent charge transfer reactions were observed. The oxidation of genistein involves the formation of two oxidation products that undergo two electrons and two protons reversible redox reactions. The electrochemical behavior of biochanin A and daidzein, two genistein chemical analogues were also investigated to identify genistein electroactive centers. Both compounds undergo two consecutive, irreversible, and pH-dependent redox reactions. the potential peak potential is influenced by the position of the hydroxyl groups in their structures [62].

4.4.3 N-Containing Compounds

Alkaloids contain nitrogen in their cyclic organic compounds, but they are rarely present in nature. These are mostly soluble in aqueous solutions; thus, they are conveniently extracted in water upon the protonation of the nitrogen. This group has some of the most famous and infamous compounds, such as cocaine, caffeine, nicotine, and morphine, which are known for their anxiolytic, analgesic, and hallucinogenic effects, and often have physiological effects on the central nervous system. Although it is a small group of metabolites, 50% of plant-derived pharmaceuticals are alkaloids [73].

4.4.3.1 Cocaine

Cocaine is one of the most commonly abused stimulants worldwide, produced from the leaves of a plant called coca. A research fulfilled in 2018 showed that approximately 19 million persons used cocaine worldwide (0.4% of people aged 18 to 64).

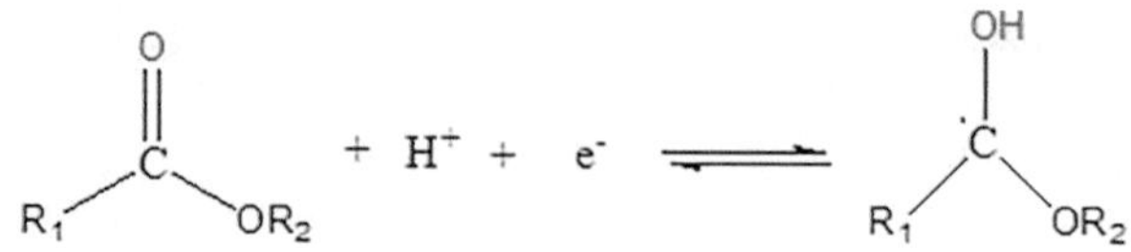

FIGURE 4.9 Electrochemical oxidation reaction of cocaine.

When cocaine is consumed at regular intervals, it creates a severe addiction that can cause life-threatening diseases. Cocaine overdose causes tachycardia and fatal consequences, such as hallucination, paranoia, hyperthermia, arrhythmia, anxiety, tremor, and heart attack. Although cocaine is metabolized to benzoylecgonine, ecgonine, and ecgonine methyl ester, respectively, in the blood circulation after consumption, it can be detectable in other biological fluids, such as blood, urine, and saliva, for 2–3 days later [74].

The electrochemical sensing process of cocaine is directly related to the elucidation of electrochemical oxidation reactions. The similarity of the responses of the bare and the cocaine contaminated electrodes can be explained by the catalytic effect of the surface semiquinones, or other oxygen containing groups. It may be assumed that the electroactive centers in the cocaine molecule are the ester groups, which are in close contact with the graphite surface:

It is further assumed that the radical C-OH appearing as the product of the C=O group reduction is stabilized by the interaction of its free electron with p-electrons of the graphite structure [75].

4.4.3.2 Caffeine

Caffeine (1,3,7-trimethylxantine, CF) is a natural alkaloid belong to N-methyl derivatives of xanthine. Beverages such as coffee, Coca-Cola, and tea are the common sources of caffeine. Physiological effects, such as gastric acid secretion, diuresis, and stimulation of the central nervous system. Caffeine is used therapeutically in combination with nonsteroidal anti-inflammatory drugs for analgesic purposes. However, high amounts of caffeine can cause untoward effects, like trembling, nausea, agitation, seizures and mutation effects such as inhibition of DNA [76].

The mechanism of electrochemical oxidation of caffeine is shown in Figure 4.10, as reported earlier 53. Substituted uric acid is firstly formed by the oxidation of two electrons and two protons at C-8-N-9 position, and secondly by the formation of 4,5 diol of uric acid analogue through the primary two electrons and two proton oxidation [77].

4.4.3.3 Nicotine

Nicotine (3–1-methyl-2-pyrrolidinyl) is an alkaloid compound present in tobacco leaves, and is a heavily used as an addictive substance among adults. As stated by the Centers for Disease Control and Prevention (CDC), about half a million Americans die each year due to exposure to secondhand smoke or smoking. Smokers burn

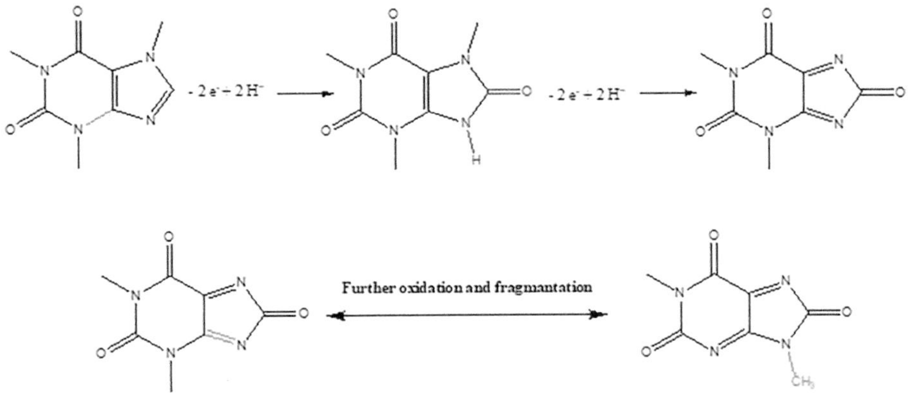

FIGURE 4.10 The mechanism of electrochemical oxidation of caffeine.

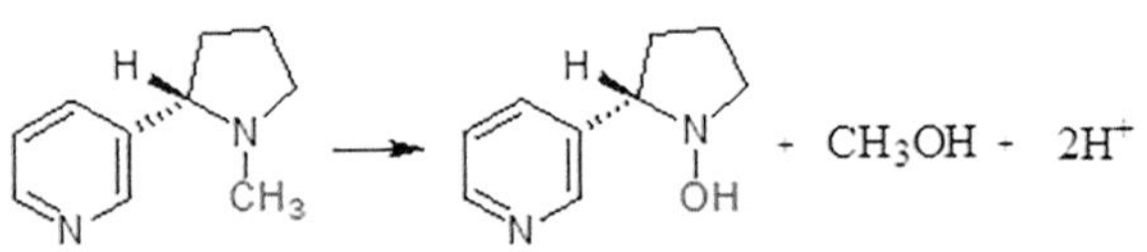

FIGURE 4.11 The oxidation mechanism of nicotine.

tobacco leaves, which causes many problems, such as heart disease, stroke, cancer, and lung diseases, and also negatively affects the human brain [78].

two electrons were participated in the oxidation process of NIC. So the expected mechanism (Figure 4.11) for the oxidation of NIC at the GO/Nq/GCE in alkaline media involves substitution of CH_3 group by OH at the nitrogen of pyrrolidine ring with formation of methanol by two protons and two electrons transfer [79].

4.4.3.4 Morphine

Morphine, (5α,6α)-7,8-didehydro-4,5-epoxy-17-methyl morphinan-3,6-diol as the chief alkaloid of opium, is clinically used to alleviate pain. Additionally, it is as a doping drug to relieve the pain caused by intense exercise activities, making athletic proficiency be decreased. According to some fulfilled research findings, using morphine as a painkiller (which is legal) in the long term, for exercise activities causes dependency in users, but the counterparts also reported reduced pain after taking saline injection (placebo) on the day of competition [80].

The electro-oxidation mechanism of morphine is shown in Figure 4.12. The oxidation peak is contributed to the oxidation reaction of the phenolic group (–OH) at the 3-position which involves one-electron transfer and is responsible for the major peak. The oxidation of the phenolic group leads to the formation of pseudomorphine (PM) as the main product. Since the structure of pseudomorphine possesses two phenolic groups, it facilitates more oxidation process. Therefore, the oxidation peak

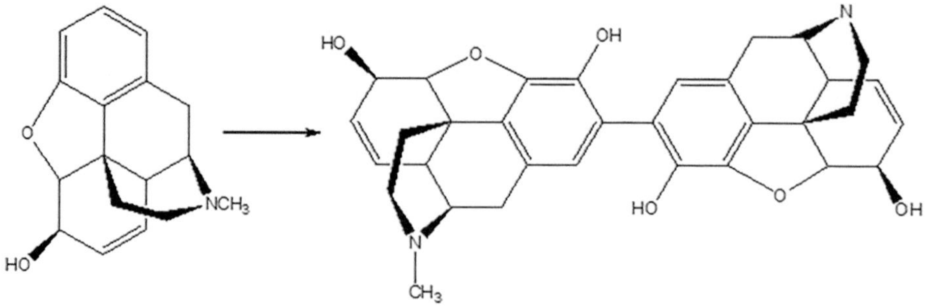

MOH PMOH

$$\text{M-OH} \rightleftharpoons \text{M-O}\cdot + e^- + H^+$$

$$\text{PM-O}\cdot + e^- + H^+ \rightleftharpoons \text{PM-OH}$$

FIGURE 4.12 The electro-oxidation mechanism of morphine.

is ascribed to the oxidation of the phenolic groups in morphine (M-OH) and pseudomorphine according to following mechanism [81].

4.4.4 Terpenes

Terpenes are a large metabolite group with different structures; up to now, more than 35,000 different terpenes have discovered. Terpenes have isoprene unites in their structure and can be be modified by cyclization reactions to make them recognizable. They are categorized into different groups based on the number of isoprene units in the carbon Skeleton. Terpenoids include metabolites with anti-tuberculosis, cancerolytic activities, anxiolytic, and mutagenic active molecules. Terpenes have no redox behavior and cannot be analyzed with electrochemical methods [73].

4.5 PREPARATION OF VOLTAMMETRIC SOLUTIONS

4.5.1 Stock Solutions of Natural Compounds

Stock solutions of natural compounds with a concentration of 0.01 M are prepared by dissolving a specific amount of the respective substances in deionized (DI) water. Next, a specific volume of drug solution is diluted with 10 mL of buffer in the electrochemical cell for electrochemical measurement and investigation.

4.5.2 Buffer Solution

In order for the flow resulting from oxidation and reduction to be only related to the amount of natural compounds in the solution, and the concentration of these species can be found through the flow, the system must be under diffusion control. For this purpose, the currents resulting from the migration and convection of the sample in the solution should be removed, respectively, by preventing the mechanical stirring of the solution and adding an inert electrolyte (buffer). For this reason, buffer is used as an inert electrolyte in electrochemical tests. Also, the reason for using different buffer pH in electrochemical tests is to find the best redox peak related to the studied natural compound, because every natural compound has a better redox peak at a certain pH. In the following, all electrochemical tests are performed in this optimal pH buffer.

- *Phosphate buffer solution (PBS).* To prepare phosphate buffer solution (PBS) with a concentration of 0.1 M, from 85% phosphoric acid solution and NaOH solution with a concentration of 0.1 M to adjust different pH values (from 3.0 to 9.0) using a pH device meter, is used.
- *Britton–Robinson buffer.* Britton–Robinson buffer is also prepared by adding appropriate amounts of 0.2 M sodium hydroxide into solutions containing 0.04 M of each of boric acid, acetic acid, and phosphoric acid.

4.5.3 Preparation of Real Samples

To measure natural compounds in real samples, including drugs and herbs, due to the complexity of the matrix in these samples, the standard addition method is used. In this method, due to the complexity of the matrix of the unknown sample, it is not possible to make a standard sample with the same matrix as the unknown sample. For this reason, the unknown sample is used to make the matrix of the samples the same. The standard addition method is divided into single-point and multi-point categories based on the use of a standard sample or several standard samples. The single-point standard addition method is not accurate enough for the range of concentrations. Therefore, the multi-point standard addition method is mostly used. For this purpose, each time a specific volume of the unknown sample is added to different volumes of the standard sample and the desired signal is recorded. Next, the relevant calibration diagram is drawn, and the concentration of the unknown sample is measured. The most important feature of the standard addition method is that it can be used to obtain the concentration of the analyte in the sample with a complex matrix.

Measurement of natural compounds by voltammetric method has advantages, such as convenience, high speed, no need for solvent, high selectivity, high sensitivity, low detection limit, and low cost, compared to other analysis methods. To measure natural compounds in real samples, including drugs and plants, by voltammetric method, real samples must be prepared first. The preparation method of these real samples is presented as follows.

- *Pharmaceutical samples.* To show the efficiency of the method for the analysis of pharmaceutical samples, the tablet sample of natural compounds is also used. After a number of tablets are completely ground and homogenized, an appropriate weight of the sample is dissolved in distilled water.

For better dissolution, the dissolution method using ultrasonic waves is used. Then, the sample is filtered by filter paper, and the solution is diluted with a ratio of 1:9 by buffer. Finally, the prepared solutions are transferred to the electrochemical cell to determine their exact concentration using the differential pulse voltammetry method.

- *Medicinal plant samples.* These samples are washed and cleaned with water, and after crushing and filtering, their extract is extracted by a solvent. The obtained extract is continuously shaken for several hours. The extract is filtered, and then the solvent is evaporated. The remaining mass is diluted in PBS with a ratio of 1:9. Finally, the prepared solutions are transferred to the electrochemical cell to determine their exact concentration using the differential pulse voltammetry method.

4.6 MANUFACTURE OF WORKING ELECTRODES FOR VOLTAMMETRY MEASUREMENT

4.6.1 Carbon Paste Electrode (CPE)

One of the suitable substrates used in electrochemical sensors is carbon paste, which is often prepared by mixing graphite powder and non-conductive adhesives, such as silicone oil or paraffin. Carbon paste electrodes have unique properties, including a wide potential window (from +1.3 to −1.4 V, compared to the reference electrode), low background current, good stability, easy preparation, low cost, simplicity, and no memory effects due to the ability or the renewability of the surface. These properties have led to the use of these electrodes in many fields, including the measurement of natural compounds.

The working electrode is prepared with graphite powder and synthesized modifiers. First, an amount of 1–10% by weight of the modifier powder is added to the graphite powder, and it is dispersed in ethanol by an ultrasonic bath for 5 minutes. It can be the obtained homogeneous paste is pressed at the end of a glass tube and the electrical connection is made by placing a copper wire in the tube. To obtain suitable and renewable surfaces, the electrode is polished with paper each time [13, 14, 82–87].

4.6.2 Glassy Carbon Electrode (GCE)

To make the modified glassy carbon electrode, the surface of the electrode must be cleaned first. To clean the electrode, GCE is carefully polished with alumina (0.05 μm) on the polishing cloth to make its surface smooth and polished. Then it is washed twice with distilled water and dried in an oven to completely wash the remaining alumina particles on the surface of the electrode. Then, about 1–5 microliters of the modifier suspension is placed on the surface of the cleaned electrode using a micropipette. Finally, the electrode is placed inside the oven so that the desired solvent evaporates and the modifier is placed on the electrode. Modifying the electrode surface with this method increases the conductivity of the glassy carbon electrode surface [19–21, 88–104].